URANIUM ENRICHMENT AND PUBLIC POLICY

AEI-Hoover
policy studies

The studies in this series are issued jointly
by the American Enterprise Institute
for Public Policy Research and the Hoover
Institution on War, Revolution and Peace.
They are designed to focus on
policy problems of current and future interest,
to set forth the factors underlying
these problems and to evaluate
courses of action available to policy makers.
The views expressed in these studies
are those of the authors and do not necessarily
reflect the views of the staff, officers,
or members of the governing boards of
AEI or the Hoover Institution.

URANIUM ENRICHMENT AND PUBLIC POLICY

Thomas Gale Moore

American Enterprise Institute for Public Policy Research
Washington, D.C.

Hoover Institution on War, Revolution and Peace
Stanford University, Stanford, California

AEI-Hoover policy studies 25

Library of Congress Cataloging in Publication Data

Moore, Thomas Gale.
    Uranium enrichment and public policy.

    (AEI-Hoover policy studies; 25) (Hoover Institution studies; 62)
    1. Atomic energy industries—United States. 2. Uranium—Isotopes. 3. Atomic power-plants—United States. I. Title. II. Series. III. Series: Hoover Institution studies; 62.
HD9698. U52M65 1978   338.4'7'6201892931   78-2440   ISBN 0-8447-3286-9

Printed in the United States of America

Contents

INTRODUCTION 1

1 THE NUCLEAR FUEL CYCLE 3

Natural Uranium   4
Enrichment   5
Nuclear Reactors   5
Radioactive Wastes   6
Summary   6

2 THE ENRICHMENT PROCESS 9

Gaseous Diffusion   9
Tails Assay   11
Gas Centrifuge   14
Other Enrichment Processes   15
Summary   16

3 THE DEMAND FOR ENRICHMENT 17

The Demand for Electricity   17
Nuclear Power's Share: Determining Factors   19
Nuclear Power's Share: Demand and Capacity Projections   27
The Demand for Separative Work   29
Summary   30

4 A HISTORICAL PERSPECTIVE 33

Reactions to the Bill   38
Other Effects of the Bill   41

5 URANIUM-ENRICHMENT OPTIONS 45

    The Carter Administration Plan   45
    A Government Corporation   49
    ''Privatization'' of New Capacity   50
    Full ''Privatization''   53

6 POLICY PRESCRIPTIONS 57

    APPENDIX: The Plutonium Recycle Question 61

Introduction

Enriched uranium, necessary under current conditions to generate electricity in nuclear reactors, is now a government monopoly: all of the existing enrichment plants are government owned, having been built to enrich uranium for military purposes. Civilian demand for enriched uranium is growing, however, and will be dominated by the electrical power industry. The basic policy issue is who is to own, construct, and operate the needed additional capacity. The Ford administration proposed that the private sector be encouraged to do so through numerous guarantees. Alternative policies include continued government monopoly and operation by the Department of Energy, establishment of a government corporation, and complete relinquishment of government control, with the private sector taking the risks of constructing additional capacity on its own. An analysis of these alternatives is the subject of this monograph.

Chapter 1 describes the fuel cycle and the public policy issues involved. In chapter 2 the enrichment process, current capacity, and the operation of the plants by the government are described. Chapter 3 estimates future demand for electricity in the United States and forecasts the demand for nuclear power and for enrichment services, concluding that additions to generating capacity will be nuclear and that uranium-enrichment capacity must therefore be increased.

Chapter 4 describes the historical background, including the Nuclear Fuel Assurance Act, the opposition to it, and its demise. The original Ford proposal would have established a cooperative arrangement to encourage private enterprise, with certain guarantees, to construct enrichment facilities. Uranium Enrichment Associates (UEA) wished to build a gaseous diffusion enrichment plant but after a cool congressional reaction shelved the plan in November 1976. Three private centrifuge enrichment proposals, which would have been covered by the administration's proposals, are also described.

Chapter 5 outlines the basic policy options the government faces and analyzes the advantages and disadvantages of each. The final chapter summarizes the findings and concludes that, from an economic efficiency view-

1

point, selling the existing plants and turning the industry over to the private sector would be best, although politically it may be difficult to secure congressional approval of such a policy.

An appendix describes some of the issues involved in plutonium recycling. Although the risk of reprocessing appears to have been exaggerated, it seems, nevertheless, that reprocessing spent fuel is probably only a marginally viable activity in terms of its economics, at least until such time as uranium becomes much more costly.

1

The Nuclear Fuel Cycle

The nuclear industry has come a long way since it was created to help win World War II. Atomic power has been increasingly used for commercial civilian activities. Unless frustrated by regulation, its role seems likely to grow in the future. Even before the sharp increase in oil and coal prices, nuclear power generation looked like an attractive option available for most utilities. In coal-rich parts of the country, fossil fuels are likely to continue to predominate; but in the rest of the country, nuclear power will be the most economical source of electricity.

A main contribution of atomic physics to society has been the use of nuclear energy to generate electricity. Theoretically, energy can be released from the atom through one of two processes: fission and fusion. In fission an atom breaks down into other elements of lower atomic number, releasing energy in the process. In fusion two or more atoms of low atomic number are combined to form another element of higher atomic number; this can also release energy, although it takes considerable energy to start the process.

Fusion holds the potential for the provision of almost limitless clean energy in the future. The process will combine two atoms of hydrogen, one of the earth's most plentiful elements, to form a helium atom, releasing energy in the process. But the heat necessary to start this process is so intense that researchers have been unable as yet to contain the fusion process and maintain the reaction. Until this and other problems are solved, fusion must remain only a promise for the future.

For the rest of this century and probably well into the twenty-first, nuclear power will be based on the fission process using either enriched uranium or plutonium as fuel. Neither appears naturally in nature and therefore both must be made. Natural uranium consists of the isotopes U_{238}, U_{235}, and U_{234}, of which U_{235} represents only 0.711 percent of the total and U_{234} a negligible amount. But only U_{235} is fissionable and thus useful for generating power. Plutonium is not found naturally on earth. It is a man-made element that can be

3

created from U_{238} in a nuclear power plant. It is conceptually possible to produce more fissionable material in a power plant than is used by the plant in generating electricity or energy. However, the technology for such a "breeder" reactor is not yet well enough developed to predict when it will be commercially feasible, even though the French have built a demonstration plant and in cooperation with West Germany and Italy are building a commercial-scale facility.

The current nuclear generating systems which are most widely used here and abroad are light water reactors that use uranium enriched to 2 to 4 percent U_{235}. These reactors have proved to be efficient and economical in comparison with alternative sources of power. While environmentalists have raised questions about the safety of these reactors and about nuclear power in general, such problems are outside the scope of this study. However, it is assumed here that environmental and safety considerations will not block future construction and growth of the nuclear power industry.

If the nuclear power industry is to grow and supply our economy with the much needed electrical power, all parts of the industry must expand their capacity. The generation of electrical power by nuclear reactor is a complex procedure involving numerous subindustries.

Natural Uranium

First, raw uranium (uranium oxide) must be found and mined. Through the 1960s, there was a worldwide surplus of uranium, and the United States barred the import of foreign uranium for use in domestic reactors in order to protect American uranium mining interests. Until 1974 uranium oxide sold for about $8 per pound. Increased demand since 1973, resulting in part from the increased cost of competing sources of energy, has pushed prices up, with purchases in early 1976 at prices as high as $40 per pound.[1] U.S. reserves are estimated to total between 1.8 and 3.7 million tons; the larger figure would be adequate to supply ERDA's estimates of needs through the year 2000.[2]

The rapid escalation of the price for uranium since 1973 has greatly spurred exploration, and it seems likely that major new deposits will be found both here and abroad. Some mining companies have tripled their drilling and major American oil companies are invading the area. Searches are underway in the Rocky Mountains, the Great Lakes region, and New England.[3] In Canada, too, a significant increase in mine capacity is taking place. South Africa and

---

[1] *Wall Street Journal,* June 7, 1976, p. 1.

[2] Energy Research and Development Administration (ERDA), *Resources, Fuel and Fuel Cycles and Proliferation Aspects,* ERDA 77–59 (ERDA, Washington, D.C., Assistant Administrator for Nuclear Energy, April 15, 1977), p. 9.

[3] *Wall Street Journal,* June 7, 1976, p. 1.

Australia are reactivating idle uranium mining facilities, and Niger, France, Gabon, and Sweden also plan considerable expansion in mining activity.[4]

After mining, uranium ore must be milled and converted into a uranium concentrate called yellowcake (U_3O_8). There are fourteen commercial mills operating in the United States; the yellowcake they produce is sent to one of two American plants which convert it into a gas, uranium hexafluoride (UF_6).

Enrichment

The next step is to enrich the uranium in U_{235}. As noted above, natural uranium is 99.3 percent U_{238}. The concentration of U_{235} atoms must be increased in order to provide for a sustained chain reaction. This enrichment process, which is described in detail in chapter 2, is the most costly aspect of the fuel cycle, and enrichment capacity is where the bottleneck in the nuclear power industry may arise. At present there are three enrichment plants in the United States, all owned by the government.

In the process of concentrating the U_{235} atoms, the enrichment plants produce two outputs. One output, called the "product," is uranium with 3 to 4 percent U_{235}. The other output, called the "tails," is uranium depleted in U_{235} with a concentration of only 0.2 to 0.4 percent. The enriched uranium is sent to a fabrication plant and formed into pellets which are then placed in zirconium tubes. The tubes are assembled into bundles and sent to nuclear power plants. There are seven American companies involved in the fabrication of nuclear fuel.[5]

Nuclear Reactors

In the power plant the close proximity of the concentrated U_{235} atoms produces a sustained reaction that releases heat as the atoms undergo fission. The reactor fuel consists of approximately 100 tons of uranium encased in zircaloy tubes a half-inch in diameter and about twelve feet long. There are from 50 to 200 rods in each fuel bundle and the reactor contains several hundred bundles. The large steel vessel in which the reactor fuel is kept is filled with water, which is needed both to cool the fuel and to maintain the chain reaction. The heated water forms steam which drives a turbine to generate electricity.

As the uranium undergoes fission, it forms over 100 radioactive isotopes called fission products, such as plutonium-239, iodine, krypton-85, netrium-99, cesium-137, and strontium-90. These fission products subsequently decay further and release radiation. Many of the fission products last only a few

---

[4] ERDA, *Uranium Enrichment Conference,* CONF-751134, (ERDA, Office of Public Affairs, Technical Information Center, November 1975), p. 46.

[5] ERDA, *Uranium Enrichment: A Vital New Industry,* ERDA-85, October 1975, p. 12.

5

minutes before decaying into nonradioactive forms, but others take much longer; in some cases years must pass before they decay into nonradioactive forms.

Radioactive Wastes

The crucial unsolved problem for the nuclear power industry is what to do with spent fuel—the depleted uranium plus the fission products. Currently, spent fuel is stored and cooled in a large water basin at the nuclear plants. Typically, a storage pool is capable of holding about one full core of spent fuel. For a number of nuclear plants, on-site storage capacity is rapidly being exhausted, however, so other procedures will become necessary in the future. It is possible in most cases to increase the storage capacity of each pool to two or three cores by making structural changes to increase the allowable density. This will make storage of ten years' supply of spent fuel possible.

Spokesmen for the nuclear industry have been urging the Nuclear Regulatory Commission (NRC)[6] to license the construction of chemical reprocessing plants.[7] In such plants the remaining uranium and plutonium created in the reactor could be separated from the other radioactive fission products. The remaining radioactive waste products could then be converted into a compact solid and stored indefinitely in a government repository. A number of possible long-term storage sites for waste fuel have been suggested, but the government has not yet adopted any.

The recovered uranium could be returned to the enriching plants to be put back into the fuel cycle. If the Nuclear Regulatory Commission grants the authority, the recovered plutonium can be sent to a fuel fabrication plant and made into fuel for power plants. It has been estimated that plutonium recycling could reduce enrichment requirements and yellowcake requirements by 15 percent.[8] Some of the policy issues involved in processing of spent fuel are discussed in the appendix to this monograph.

Summary

The nuclear fuel cycle consists of the mining of uranium ore, the milling of the ore into yellowcake, enrichment—the concentration of uranium-235 atoms to

[6] In 1975 the former Atomic Energy Commission was divided into the Nuclear Regulatory Commission to regulate safety aspects of nuclear power and the Energy Research and Development Administration (ERDA) to develop and promote all forms of energy including but not restricted to nuclear. The latter organization has been incorporated into the Department of Energy (DOE).

[7] See report on talk by A. Eugene Schubert, president, Allied-General Nuclear Services, as described in *Nuclear Industry,* December 1975, p. 18, and "Pu Recycle Rule Urgent," *Nuclear Industry,* July 1975, pp. 7–10.

[8] Talk by Richard A. McCormack, president, General Atomic Company Power Systems Group, as described in "Fuel Cycle Bottlenecks," *Nuclear Industry,* November 1974, p. 25.

an average level of about 3 to 4 percent of the total—the fabrication of fuel rods, the concentration of the rods in a reactor to produce a chain reaction, the removal of the spent fuel (including fission products) and its storage, and then possibly the chemical separation of the reusable uranium and plutonium from the waste products which must be stored indefinitely. The last step is not necessary to the cycle but would decrease the amount of raw uranium that must be mined.

2

The Enrichment Process

In the United States today enrichment of uranium is carried out in three government-owned plants. These plants—located in Oak Ridge, Tennessee; Paducah, Kentucky; and Portsmouth, Ohio—were built during and after World War II to produce highly enriched uranium for military purposes. Today no uranium is being enriched for bombs and only a small amount is being highly enriched (over 20 percent), mainly for the navy. Thus, these plants and any future enrichment capacity will be used entirely for the nuclear power plant fuel cycle.

Gaseous Diffusion

The three government-owned enrichment plants, which are operated, under the direction of the Department of Energy, by Union Carbide (Oak Ridge and Paducah) and Goodyear (Portsmouth), use the gaseous diffusion method of enriching. The private companies work under a cost-plus-fee contract and simply carry out the detailed policies set by the Department of Energy. In the gaseous diffusion process, uranium is converted to a gas—uranium hexafluoride, UF_6—which is extremely reactive with water and very corrosive to most common metals. To maintain uranium hexafluoride in a gaseous state, it must be heated and kept under pressure. In this state it is pumped through a tube containing a barrier through which the gas can diffuse (see figure 1). A slightly higher percentage of the lighter U_{235} atoms diffuse through the barrier than U_{238} atoms, thus producing a more enriched mixture on the far side of the barrier and leaving a slightly depleted mixture on the original side. Since a single stage of this diffusion process increases the concentration of U_{235} atoms by a factor of only 1.00429, it takes an enormous number of stages to enrich normal uranium to 3 to 4 percent U_{235}.[1] Each stage uses electric power for the pumps and to

---

[1] U.S. Atomic Energy Commission, *AEC Gaseous Diffusion Plant Operations*, ORO-658, February 1968, p. 3.

**Figure 1
A Gaseous Diffusion Enrichment Stage**

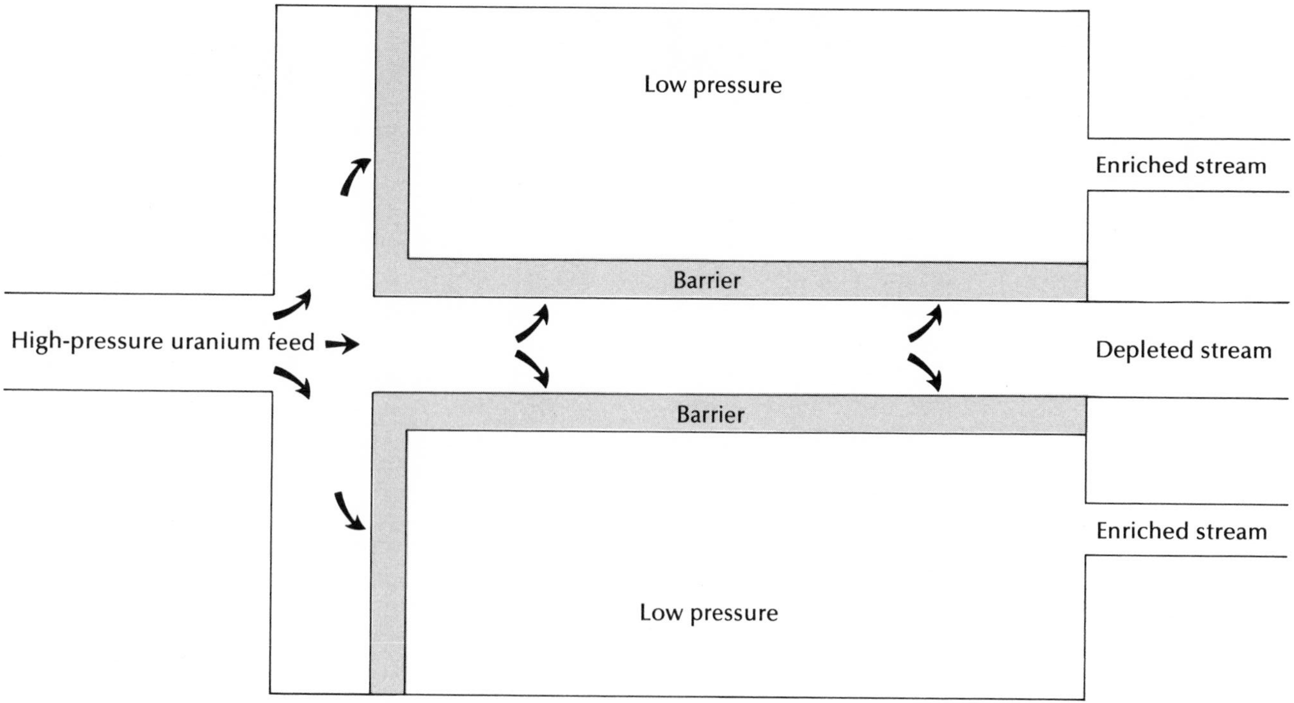

Enriched stream

Depleted stream

Enriched stream

Low pressure

Barrier

Barrier

Low pressure

High-pressure uranium feed

maintain the temperature and pressure on the feed. Thus, the enrichment process demands large quantities of electric power.

Currently the Department of Energy (DOE), which runs these plants for the U.S. government, is involved in two programs to increase the capacity of the existing gaseous diffusion plants. The Cascade Improvement Program (CIP) uses improved compressors and barriers to increase the amount of separation that a single stage can accomplish. The other program, entitled the Cascade Uprating Program (CUP), is aimed at increasing the flow of UF_6 through the cascade by raising the electrical power and improving the system. When completed near the end of 1981, these programs together will increase the capacity of the three government-owned plants by 60 percent.[2] Congress has also authorized construction of an addition to the Portsmouth facilities; the new plant will use the gas centrifuge technique, discussed in detail below.

The output of these plants is expressed in terms of "separative work units" or SWUs. The exact definition of a separative work unit is complicated and need not be explained here; suffice it to say that, if 5.479 kilograms of normal feed (0.711 percent U_{235}) are fed into an enrichment plant and an output of 1 kilogram of 3 percent U_{235} is produced, with a stream of depleted uranium ("tails") of 4.479 kilograms at an assay weight of 0.2 percent U_{235}, then it took 4.305 SWUs to accomplish this enrichment.[3]

To load a single typical 1,000-megawatt reactor for the first time takes about 200,000 SWUs, assuming a 0.3 percent tails assay; an additional 89,000 to 99,000 SWUs annually are required to keep it operating (assuming no plutonium recycling).[4] When the CIP and CUP expansion programs are completed, the capacity of the enrichment plants will expand to 27.7 million SWUs, or enough to handle the annual needs of about 300 nuclear power plants, assuming a 0.3 percent tails assay and no recycling of plutonium. However, full capacity will not be reached before 1984-1985.

Tails Assay

From the existing diffusion plants, more or fewer separative work units can be used to produce the same tonnage of enriched uranium, depending on the tails assay. If the tails assay is high, say, around 0.4 (weight percent U_{235}), then additional feed of yellowcake will be needed to produce a given tonnage of output with less separative work. In actual output it would take 52.6 percent more feed but 33.9 percent fewer separative work units to produce the same quantity of 3 percent enriched uranium at 0.4 percent tails assay rather than at

[2] ERDA, *Uranium Enrichment Conference,* pp. 93, 97.

[3] U.S. Atomic Energy Commission, *Selected Background Information on Uranium Enriching,* ORO-668, March 1969, p. 5.

[4] U.S. Congress, House, Subcommittee on Energy of the Committee on Science and Astronautics, *Energy Facts,* 93d Congress, 1st session, November 1973.

0.2 percent tails assay. To put it another way, given the existing diffusion plants, 51.2 percent more output can be produced in a given period of time when the plants are operated with a 0.4 percent tails assay than when they are operated at a 0.2 tails assay, but 130.7 percent more raw uranium is required.[5]

From these figures it is obvious that the capacity of a given plant in terms of the tonnage of enriched uranium produced is not fixed. Output depends on the tails assay. The optimum tails assay depends on the price of raw uranium and the cost of enrichment. The more expensive yellowcake is, the lower is the optimum tails assay, given the price of separative work units. Alternatively, the more output needed from a plant during a given period, the higher the tails assay will be and the more expensive will be the enrichment because of the additional natural uranium that will be needed.

Figure 2 shows the relationship between the cost of separative work, the cost of yellowcake, and the optimum tails assay. For example, if one SWU costs $80 and the price of yellowcake is about $30 per pound, then the tails assay should be about 0.23. On the other hand, if yellowcake is cheaper then the optimum tails assay becomes higher.

Currently DOE is operating the diffusion plants with a tails assay of 0.25 percent.[6] Plans were to increase this tails assay gradually, but with the congressional directive to build the add-on plant, DOE announced that the additional capacity will be used to lower tails assay to 0.2 to 0.25 percent U_{235} in order to conserve "our limited natural uranium resources."[7] Of course, uranium tails are stockpiled and will always be available for enriching if that becomes economically feasible.

There is little argument that the current pricing formula for enrichment services, which is based on recovery of the government's costs over a reasonable period of time, understates economic costs. Government costs do not include any return on government equity, are based on low government interest costs (risk free), and average cost pricing for electricity. This means that enrichment is subsidized and too low a tails assay is established.

If plutonium recycling is begun, considerably less enriched uranium will be needed. ERDA estimated that with plutonium recycling its domestic enrichment customers will save over 30 percent natural feed. They expect this saving to increase to over one-third by 1988.[8] ERDA's estimates of production from reserves plus potential resources would just about meet the demand for natural uranium domestically through 1987 with plutonium recycling but would fall short without it.[9]

---

[5] Calculated by the author from data formulas given in U.S. Atomic Energy Commission, *A.E.C. Gaseous Diffusion Plant Operations*, p. 37.
[6] ERDA, *Uranium Enrichment Conference*, p. 36.
[7] Letter to the author from Roger W. A. LeGassie, Assistant Administrator for Planning Analysis and Evaluation, dated October 4, 1976.
[8] ERDA, *Uranium Enrichment Conference*, p. 47.
[9] Ibid., p. 52.

Figure 2
Optimum Tails Assay

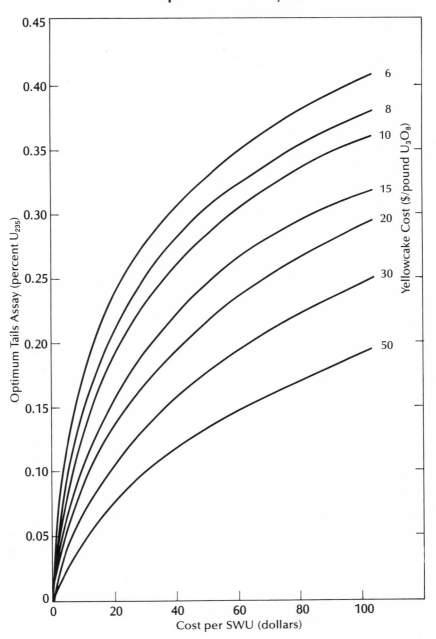

As of the fall of 1977, DOE had signed contracts to supply enriched uranium to 178 domestic utilities representing 209 reactors, which will have a capacity of 204 thousand megawatts of electricity when in operation.[10] DOE has 137 firm and 5 conditional contracts with foreign customers to supply enrichment services for 143 and 5 reactors respectively. Together, the domestic and foreign contracts which will supply 357 reactors producing 323 thousand megawatts of electricity will take all of DOE's planned output available before the new centrifuge facilities come on line in 1988. Only the construction of the additional add-on facilities will permit the Department of Energy to supply enrichment services to more power plants.

Gas Centrifuge

A number of other techniques exist to enrich uranium, although none has been put to full commercial use. The most advanced and most immediately promising is the gas centrifuge. The Carter administration has elected to make the next addition to enrichment capacity, authorized by Congress in 1976, an 8.8-million-SWU centrifuge plant at Portsmouth, Ohio.[11]

The centrifuge works like a cream separator. Uranium hexafluoride is whirled in cylinders at a high speed with the result that centrifugal force pushes the heavier atoms, U_{238}, to the outside, leaving a greater concentration of lighter ones, U_{235}, on the inside where they can be extracted.

One potential problem with this enrichment technique is reliability. A centrifuge plant with any reasonable capacity would need to have a very large number of machines. For example, a plant with a 9-million-SWU capacity might have as many as 99,000 centrifuges, each of which could fail. If the mean time between failures were three years (a 33 percent failure rate per year), an average of ninety machines would be down each day.

However, a centrifuge plant has several important advantages over a gaseous diffusion plant. For a plant of a given size the capital requirements are the same as those of a diffusion plant, but economies of scale are not important after about 3 million SWUs per year, whereas the minimum feasible diffusion plant would be three times as large. Moreover, energy requirements are about 10 percent of those for diffusion plants.[12]

Since 1940 the government has been developing the gas centrifuge process. An experimental facility has been constructed and cascade operation began in January 1977. As pointed out above, a significant number of the centrifuge

[10] Ted Aukrum, Chief of Commercial Policy, Uranium Enrichment Division, Department of Energy, November 7, 1977.

[11] *Nuclear Industry,* August 1977, p. 14.

[12] Comptroller General of the U.S., *Evaluation of the Administration's Proposal for Government Assistance to Private Uranium Enrichment Groups,* Report to Joint Committee on Atomic Energy, U.S. Congress, RCD-76-36, October 31, 1975, p. 5.

machines could be down at any one time unless they are extremely reliable. However, as one expert predicts, "if the life of a gas centrifuge is long and maintenance costs sufficiently low, this process could produce separative work at a lower cost than any other processes."[13]

In 1971 Britain, the Netherlands, and West Germany formed a consortium called URENCO to develop and build a commercial gas centrifuge plant. A Dutch plant began operating in 1972 and a German pilot plant located next to the Dutch one reached full operation at the end of 1973. These plants have been operated at close to capacity by chemical plant operatives and unskilled workers. Failures during the first 100 days were less than 2 percent and for the subsequent period the overall average has been reported to be 2 percent.[14] This indicates a failure rate for new machines of about 7 percent, but with aging the rate can be expected to increase. The consortium is planning to invest $2.5 billion in the construction of a 10,000-million-SWU plant to be in operation by 1985. Construction lead times are four to five years before full operation.

Other Enrichment Processes

Two or more other enrichment processes are under development. West Germany is experimenting with an aerodynamic nozzle method while South Africa has a secret helikon cascade method. Both of these approaches are said to require greater energy consumption than gaseous diffusion with capital costs in the same range. Where energy costs are low enough, the nozzle process might be in the same cost range as gaseous diffusion.[15] South Africa is completing a pilot plant based on its technology and plans commercial production.

Probably the process with the greatest potential, although it is at the earliest stages of development, is the laser. Very sophisticated lasers are used to produce a "chemical separation" recovering 96 percent of the U_{235} from natural uranium and leaving 0.03 percent tails.[16] If successful this technique should separate the uranium isotopes at very low cost compared to other techniques. While several variants of the laser approach are being worked on, the Lawrence Livermore Laboratory in Lawrence, California, has used one to enrich a minute quantity of natural uranium to better than 60 percent U_{235}. Their process used a laser to excite only the electrons of U_{235} atoms of a thin stream of natural uranium vapor to an energetic level. Ultraviolet light kicked some of the electrons free, leaving the U_{235} atoms positively charged. A negatively charged collector cup then captured them.[17]

[13] Manson Benedict in *Nuclear Industry,* July 1976, p. 6.

[14] David Smith, "URENCO Pushes Market Plan," *Nuclear Industry,* March 1976, p. 32.

[15] Manson Benedict in *Nuclear Industry,* July 1976, p. 6.

[16] *Nuclear Industry,* January 1974, p. 18.

[17] Ibid., June 1974, p. 44.

Summary

Enrichment is the most expensive and crucial step in the nuclear fuel cycle. Except for experimental production, all enrichment to date has been carried out by the gaseous diffusion method, in which increasing the concentration to 3 or 4 percent takes a vast number of stages, the exact number depending on the tails assay. The total enrichment capacity necessary to supply power depends on the growth in the number of plants, the tails assay, and the decision on recycling plutonium. The demand created by the growth in the nuclear power industry is discussed in the next chapter.

3

The Demand for Enrichment

The demand for enrichment depends on the demand for electricity, on the proportion of the demand supplied by nuclear power, on the availability of plutonium recycling, and on the tails assay used in the enrichment process. The optimum tails assay itself depends upon the relative price of yellowcake and separative work units (SWUs). In this chapter we look at each of these factors in turn.

The Demand for Electricity

Historically the demand for electricity in the United States has grown about 7.7 percent annually.[1] In the last two years, partly because of the sharpest recession in postwar history and partly because of the rapid rise of energy costs, electricity sales have not grown at all. Both of the factors that have held down electricity growth, however, are temporary. The economy seems to have at least partially recovered. If this trend continues, there should be a sharp growth in the economy and in electricity demand. As for the higher fuel prices that resulted in part from the Organization of Petroleum Exporting Countries (OPEC) cartel and in part from the devaluation of the dollar, they did significantly reduce the quantity of electricity bought. However, once consumers adjust to the new relative prices, demand should resume its typical growth pattern. While oil and coal prices increased very sharply between 1973 and 1975, the rate of increase has slowed and during 1976 it leveled out for coal. In recent months prices have again moved upward.

Increasing environmental costs will have to be reflected in higher prices. In addition, as old, low-interest-rate issues are rolled over, interest costs of utilities will mount. These factors mean that there will still be upward pressure

---

[1] *Statistical Abstract of the U.S., 1975* (Washington, D.C.: U.S. Bureau of the Census, 1975) table 889, p. 536.

Table 1

Projected U.S. Installed Electricity Capacity, 1980–2000
(in thousands of megawatts)

Year	Annual Growth Rate			
	5%	6%	7%	8%
1975 (actual)	505	505	505	505
1980 projected	645	676	708	732
1985 projected	823	904	993	1,060
1990 projected	1,050	1,210	1,393	1,536
1995 projected	1,340	1,620	1,954	2,226
2000 projected	1,710	2,167	2,741	3,226

Source: 1974 and 1975 data from *Statistical Yearbook of the Electric Utilities Industry for 1975,* Edison Electric Institute.

on prices in the future. Nevertheless, price increases are not likely to be as fast or as dramatic as in the 1973-1975 period.

The ecology movement, the promotion of energy saving, and the smaller family size, bringing slower population growth, should all reduce the growth rate of demand for electricity. Nevertheless, the population will continue to grow throughout this century. Most energy-saving steps have already been taken in the last year or two and will have the effect of a one-time reduction in demand. Finally, it seems unlikely to this author that ecologists will succeed in convincing significant numbers of Americans to give up their notions of the "good life." Therefore, a reasonable guess as to the future demand for electric power is that it will grow a little more slowly than it has traditionally. Table 1 projects the growth in installed electric capacity under various assumptions.

Though the general trend is clear, there is considerable uncertainty about the future rate of growth of electric power demand. In December 1976, the Federal Power Commission concluded: "While not unanimous, the consensus of projections indicates an average growth rate during 1976-1985 between the historic 7 percent on the high side to 4 percent on the low side."[2] The revised forecasts issued by the Bureau of Mines predicted an annual growth rate from 1974 to 2000 of 5.5 percent.[3] Thus, a reasonably conservative estimate would be that electrical generating capacity will grow 5 percent annually.

The slower growth in demand is predicated on the assumption that relative energy prices will not resume the steady decline that they followed during the

---

[2] The Bureau of Power, Federal Power Commission, *Factors Affecting the Electric Power Supply 1980-85: Executive Summary and Recommendations,* December 1, 1976, p. 8.

[3] W. G. Dupree, Jr. and J. S. Corsentino, *U.S. Energy through the Year 2000* (revised) (Washington, D.C.: Bureau of Mines, Department of Interior, 1975), table 15.

post–World War II period.[4] This decline in prices for electricity resulted from a fall in the price of fuel and an increase in the productivity of electricity-generating plants. While it is possible, in fact even reasonable, to believe that fuel prices will decline in real terms over the next few decades, the available evidence suggests that future gains in productivity in electric generation may be small or nonexistent. Thus, prices of electricity will fall only slightly, if at all, in the future, with the consequence that the growth in demand will be mainly a function of the growth in overall economic activity.

Nuclear Power's Share: Determining Factors

The share of the growth in electrical-generating capacity that is produced by nuclear power depends on four major factors that are difficult to predict. One crucial variable is the relative costs of generation of electricity by nuclear power plants and by fossil fuel plants. A second is the reliability of the nuclear plants. The third imponderable is the environmental costs of nuclear plants and fossil fuel plants. Fourth is the political environment, which will reflect the perceived degree of safety of nuclear plants. Each of these factors merits closer examination.

Relative Costs. The relative costs of generating electricity by nuclear or fossil fuel plants depend on the cost of fuel—oil, gas, coal, and uranium—the cost of constructing nuclear plants (including the time necessary to get the plants on the line), and the costs of fossil fuel plants. Prices of the fossil fuels, as fairly close substitutes, tend roughly to move together, at least when government regulations permit. Since 1973 coal prices have risen sharply, reaching a peak in about December 1974 and then falling about 15 percent in the next year and continuing to edge lower through most of 1976.[5] On the other hand, crude oil prices increased through 1975, fell only with the elimination of the fee the federal government had imposed on oil imports and with the introduction of controls on prices of "new" oil, and then resumed their climb.[6] In February 1977 they were 13 percent above the previous year's level. The rise in the price of crude oil within the United States during 1976 was due mainly to the increase in the proportion of oil imported at world prices and of "new" oil sold at world prices. With old oil prices fixed at $5.25 per barrel and world prices being more than twice as high, the change in proportion was bound to raise the average domestic wholesale price of crude oil, even if the world price was not rising.

This rise in the price of oil is not surprising. The tendency will be for the

---

[4] Edward J. Mitchell, *U.S. Energy Policy: A Primer* (Washington, D.C.: American Enterprise Institute, 1974), p. 12.

[5] Computed from prices listed in *Survey of Current Business,* May and October 1976, and March 1977.

[6] Ibid.

world price of oil to move toward the cost of developing the marginal barrel of oil. Over time, as the cheapest and easiest deposits of oil are exploited, deposits that are more expensive to develop must be used. Prior to 1970 this natural tendency appears to have been offset by improvements in technology that reduced the price of finding and extracting oil faster than the increase in the cost due to the use of less available deposits. Since 1970 oil firms have been moving into areas such as the North Slope of Alaska or the North Sea with significantly higher extraction costs.

If the price of imported oil is extrapolated at the rate of increase in cost per barrel from 1970 to 1973 and adjusted for inflation, a barrel of imported oil would have sold in April 1977 for about $6 rather than the $3.41 of 1973 or the actual price of over $11.[7] If the price of oil had just kept step with inflation since 1955, it would have sold for about $6.21 a barrel. It should be noted that between 1974 and 1975 the price of crude petroleum from Venezuela rose by 5 percent while consumer prices generally went up to 9 percent.[8] Basically, crude oil imported prices have roughly kept even with inflation or lagged since the OPEC cartel raised prices sharply in 1973. Thus, there is no reason to believe that oil prices will increase significantly faster than inflation in the future.

On the other hand, real coal prices have been rising since 1950, albeit slowly prior to 1970 and the Coal Mine Safety Act.[9] Since then they have jumped sharply; in real terms they doubled between 1973 and 1974. Subsequently they have eased; projecting the costs in real terms on the basis of the trends from 1969 to 1973, coal prices should decline a bit more before leveling out. By the end of 1976 they had stopped falling and started to rise somewhat.[10]

In contrast, nuclear power appears to be cheaper than fossil power. A comparison of fifteen new nuclear power plants with sixty-six fossil plants (which started operating between 1968 and 1972) showed that the average operating cost of a fossil fuel plant was 4.24 mills per kilowatt hour, while the average for a nuclear plant was 3.43, nearly 20 percent cheaper.[11] With fuel costs accounting for 60 to 90 percent of the operating costs of generating electricity in fossil fuel plants, it would take a drop of about 25 percent in coal or oil prices to make fossil fuel power generally competitive with nuclear. In fact, since 1972 the prices of fossil fuels have increased more rapidly than nuclear fuel costs, increasing the differential.

Recently the price of uranium oxide (yellowcake) has risen quite sharply from a level below $10 per pound to an average price of $19 for delivery in

[7] *Statistical Abstract of the U.S., 1976,* table 707, p. 438. Prices adjusted using the Consumer Price Index.

[8] Ibid., p. 438.

[9] *Statistical Abstract of the U.S., 1975,* table 686, p. 421; and *Statistical Abstract of the U.S., 1964,* p. 486, p. 360.

[10] *Survey of Current Business,* March 1977.

[11] Federal Power Commission, *Steam Electric Plant Construction Costs and Annual Production Expenses, 1972,* (Washington, D.C.: U.S. Government Printing Office, 1974).

1982, with some prices over \$40.[12] If the price of yellowcake continues to escalate rapidly, the relative advantage of nuclear power will diminish, but this seems unlikely. Throughout the 1960s, yellowcake prices declined.[13] We can expect that after new supplies come on the market the price will ease again. Moreover, outlays for yellowcake compose only about one-third of the fuel costs of nuclear power, with fuel costs in turn composing only about one-quarter of total costs.[14] Thus, even marked increases in uranium fuel prices will not have a major impact on the overall advantages of nuclear power.

A utility contemplating expansion of its capacity must consider not only operating costs but also relative construction costs. Nuclear power plants on the average cost about 30 percent more per kilowatt of capacity to construct than do fossil fuel plants.[15] The Nuclear Energy Policy Study Group (NEPSG) estimated in 1977 that the capital costs of making a 1,150 MWe power plant operational by 1985 would be about \$667 per kilowatt, while the capital costs of a high-sulfur coal-fired plant with scrubbers would range from \$465 to \$743 per kilowatt.[16] Plants using low-sulfur coal would have lower capital costs. Summarizing its findings, NEPSG concluded that the relative costs of nuclear and coal power plants will depend on the part of the country. New England, Florida, and the Far West will have to pay more for the coal because of transportation costs. The South, having lower construction costs, will tend to find nuclear power more advantageous. The central and west coastal states should have relatively cheap access to low-sulfur coal from Montana and Wyoming and thus favor coal over nuclear power. Table 2 reproduces table 3-3 from the Report of the Nuclear Energy Policy Study Group and shows the relative costs of coal and nuclear power in the Midwest.

On the other hand the Electric Power Research Institute concludes a study comparing the cost of coal with nuclear generation by stating:

- Both coal and nuclear generation can be economically attractive in all regions, depending on specific circumstances.
- Nuclear generation shows an average cost advantage in all regions.
- Average cost positions of coal and nuclear generation are closer in the western part of the country, reflecting the generally lower cost of coal.
- Site specific considerations will have a major effect on cost comparisons

---

[12] ERDA, Office of Public Affairs, Technical Information Center, *Uranium Enrichment Conference,* Oak Ridge, Tennessee, November 11, 1975, CONF-751134, p. 45.

[13] *Statistical Abstract of the U.S., 1975,* table 1,153, p. 677.

[14] U.S. Congress, House, Subcommittee on Energy Research, Development and Demonstration of the Committee on Science and Technology, *Energy Facts II,* 94th Congress, 1st session, August 1975, p. 259.

[15] Computed from sample of new plants drawn from Federal Power Commission, *Steam-Electric Plant,* 1972.

[16] *Nuclear Power Issues and Choices,* Report of the Nuclear Energy Policy Study Group, Spurgeon M. Keeny, Jr., Chairman (Cambridge, Mass.: Ballinger, 1977).

Table 2
Projected Cost of Generating Electricity in the Midwest in 1985
(mills/kwh in mid-1986 dollars)

	Nuclear		Coal with Scrubbers	Coal without Scrubbers
Capital charges[a]	16.5		13.7	10.3
Operations and main-				
tenance	2.0		2.8	1.6
Fuel[b]	5.4			
U_{308}	2.5	Coal at mine	10.0[c]	4.0[d]
Conversion	0.1	Transportation	2.0[e]	11.3[f]
Enrichment	2.0			
Fabrication	0.4			
Spent fuel storage				
and disposal	0.4_{+5}			$27.2^{\pm3}$
TOTAL[g]	23.9_{-4}		$28.5^{\pm4}$	

Note: In computing these costs it is assumed that the operations, maintenance, and fuel costs remain the same (in constant dollars) over time. However, capacity factors will change with time, which has the effect of changing capital charge costs per kwh with time.

[a]Based on costs for nuclear, coal with scrubbers of $1,000, $833, and $698 per kw respectively (in 1985 dollars), equivalent to $667, $555, and $465 per kw (in 1976 dollars); on a capital charge rate of 13 percent, and on capacity factors of 60 percent for the nuclear and coal with scrubber cases and 67 percent for the coal without scrubber case.

[b]Based on tails assay of 0.2 percent for uranium-235; no reprocessing of spent fuel; costs (1976 dollars) of $30/lb. for U_{308}, $4.44/kg for conversion, $80/kg/SWU for enrichment, and $90/kg (1976 dollars) for fabrication; and on carrying charges of 28 percent for U_{308} and conversion costs, and 24 percent for enrichment and fabrication costs. In the event of reprocessing, credits for recovered uranium and/or plutonium could offset part or all of the costs of the "back end" of the fuel cycle. With reprocessing and recycling, the figure for the back end of the fuel cycle is likely to range from a cost of 0.6 mills/kwh to a credit of 0.4 mills/kwh.

[c]Northern or Central Appalachian coal, $1.08/106 BTU at the mine (1976 dollars).

[d]Montana-Wyoming coal at $0.43/106 BTU at the mine (1976 dollars).

[e]Movement of 300 miles in the East.

[f]Movement of 1,400 miles from Montana to the Midwest. These figures may suggest that movement of western coal is more expensive than movement of eastern coal. The cost of transportation, by rail, measured in *dollars per ton*, is actually less in the West than in the East, but the heat content of western coal is only about 75 percent of that of Appalachian coal.

[g]These are bus bar costs at base-load generating plants. The charge to the consumer would be 50 to 200 percent higher because of distribution costs, the need for high-cost peak power, et cetera.

Source: From *Nuclear Power Issues and Choices,* Copyright 1977, The Ford Foundation. Reprinted with permission from Ballinger Publishing Company.

between coal and nuclear generation, particularly the availability of low cost coal transportation.[17]

Reliability. Besides construction and operating costs, there are questions concerning the reliability of nuclear plants. Opponents of nuclear energy have been vocal in their claims that nuclear plants are unreliable. However, the data do not bear them out. The kilowatt hours produced during a given year as a percentage of the maximum amount that could be produced is called the capacity factor and is a measure of reliability. The simple average capacity factor in 1972 for all ten major nuclear plants that came into operation between 1967 and 1970 was 64.1 percent, while the capacity factor for the fifty-seven major fossil fuel plants that came into operation during the same period was 56.0 percent, not a statistically significant difference.[18] While there was a greater variability in the percentage of kilowatts produced relative to capacity for nuclear plants in the sample, the variances for fossil fuel and nuclear plant capacity factors were not significantly different.

Other estimates of the capacity factor have shown lower levels of reliability for nuclear power plants. If the capacity factor is weighted by the average design capacity, lower average capacity factors result, since very large nuclear units have been especially unreliable.[19] The NEPSG assumed that a 60 percent capacity is realistic for future planning. For the first 6 months of 1977, the sixty-one domestic reactors in utility service operated at an average of 64 percent of maximum dependable capacity.[20]

The reliability problem appears to reflect the normal difficulties that occur when a new technology is brought into common use. Once the plants are in operation, however, they seem to do as well as major fossil fuel plants.

Environmental Effects. The relative environmental and health costs of nuclear power versus fossil fuel plants will have a major impact on whether or not regulatory authorities permit and encourage nuclear power. As is well known, there is considerable dispute as to the relative environmental effects of the two types of technology. Fossil fuel plants, except for those using natural gas, produce considerable amounts of air pollution, especially sulfur oxides and particulates. Even though nuclear plants do not pollute the atmosphere, they do degrade the environment by releasing small but measurable amounts of radioactive material, and by necessity they dissipate relatively larger amounts of heat than do fossil fuel plants. The latter problem has led regulators to require cooling towers to dissipate the heat and has resulted in considerable controversy over the impact on the environment of warming bodies of water.

[17] Electric Power Research Institute, *Coal and Nuclear Generating Costs,* EPRI PS-45-5-SR, Special Report, April 1977 (Palo Alto: Electric Power Research Institute).

[18] Sample of new plants, Federal Power Commission, *Steam-Electric Plant,* 1972.

[19] *Nuclear Power Issues and Choices,* pp. 119–20.

[20] Federal Energy Administration, *Monthly Energy Review,* August 1977, p. 39.

As for the release of radioactivity into the environment through the daily operation of plants, even the opponents of nuclear power do not treat this as a major problem. For example, J. Martin Brown, writing in support of the California Nuclear Initiative (which would have sharply restricted the growth of nuclear power) said, "so-called 'planned' releases of radioactivity from the nuclear industry are unlikely to expose the public to large increases in radiation levels."[21] The average radiation received by an individual in the United States today has been estimated to be about 207 mrem per year (a mrem is one-thousandth of a rem—the basic measure of radiation). Of this 207 mrem, 130 mrems come from natural background radiation, 73 from medical x-rays, 4 from the fallout from weapons tested in earlier years, and 0.01 from the existing nuclear power industry.[22] If the industry expands twenty-fold and reprocessing of spent fuel becomes standard, the level of radiation that could be attributed to the nuclear power industry would increase to about 1 mrem.[23] A desirable 10 percent reduction in medical x-rays would reduce average radiation exposure by more than seven times this increase.

Both fossil fuel plants and nuclear plants can affect the health of the public through the mining of the fuel and through the effect of any pollutants emitted. Lester Lave has made some estimates of the relative health costs of coal-fired plants versus nuclear plants. These figures, presented in table 3, are based on the experience in mining in the late 1960s prior to the Coal Mine Safety Act. Although the act was intended to reduce the costs of mine-related accidents and chronic diseases, apparently it has not succeeded.[24] For valuation purposes, the loss due to death is set at $300,000 for either a coal miner or an uranium miner; this value is attributed to each case of pneumoconiosis or lung cancer in a coal or uranium miner. The value of a workday lost due to illness or disability is assumed to be $50. The large variation in the costs to the public in table 3 depends on the number of people living near a power plant and the geography and weather conditions in the area. It should be noted, however, that these costs to the public can be substantial for coal-fired plants but, as pointed out above, are negligible for uranium plants. On the basis of normal operation, the cost of nuclear plants in terms of human lives and health is much lower than that of the major alternative—coal-fired plants.

Safety. The final factor that might inhibit the growth of nuclear power is the safety question: How vulnerable are nuclear power plants to the threat of core melt, sabotage, terrorist activities, and natural disasters that might destroy a

---

[21] W. C. Reynold, ed., *The California Nuclear Initiative: Analysis and Discussion of the Issues* (Stanford: Institute for Energy Studies, Stanford University, April 1976), p. 138.

[22] Ibid.

[23] Ibid.

[24] David Henderson, *The Coal Mine Safety Act* (Washington, D.C.: American Enterprise Institute, forthcoming).

Table 3

**Dollar Estimates of Health Effects
of 1,000 KWH of Electric Generation by
Coal and Uranium Plants**

	Coal Plants	Uranium Plants
Mine accidents	8¢	0.6¢
Mine chronic diseases	24¢	1.8¢
Public costs	10¢ to $5.00	0.0003 to 0.03¢
Total	40¢ to $5.32	2.4¢

Source: Lester B. Lave, "Coal or Nuclear: The Unintended Consequences of Electricity Generation," in *California Energy: The Economic Factors* (San Francisco: Federal Reserve Bank of San Francisco, 1976), p. 176.

power plant or a reactor's shield? There has been tremendous controversy over this aspect of nuclear power. The California Nuclear Initiative and initiatives in six other states, including environmentally sensitive ones such as Oregon and Colorado, put this issue to the public in referenda. In all cases the public upheld by large majorities continued use of nuclear power.

A nuclear plant cannot explode like a bomb, but if the cooling of its core fails for some reason it can heat up and melt through its container and the concrete floor below it. Such a scenario could result in the release into the atmosphere of radioactive gases which if blown by the wind into a populated area could lead to many deaths. In addition, nuclear power plants produce spent fuel which is highly radioactive. These wastes must be either processed for recycling or stored virtually indefinitely. Even if reprocessed, some radioactive residue will remain that must be stored for centuries.

In 1972 the Atomic Energy Commission (AEC) sponsored a study on the safety of reactors that was published in final form in October of 1975.[25] While the AEC was not unbiased on the safety of nuclear power, the study was conducted by a reputable scientist, Professor Norman C. Rasmussen, of the Massachusetts Institute of Technology. Before the final report was issued, comments were requested and the views of other government agencies, environmental groups, and other interested parties were considered. Since there have been no fatalities connected with the civilian nuclear power program, the study had to estimate the risk indirectly. Basically, the finding of the report was that the risk of a fatality from a nuclear power plant was over 2,000 times smaller than the risk of accidental death from man-made causes such as traffic

---

[25] U.S. Nuclear Regulatory Commission, *Reactor Safety Study: An Assessment of Accident Risks in the U.S. Commercial Nuclear Power Plants: Main Report,* Wash-1400 (NUREG-75/014), October 1975.

accidents and firearms, or natural disasters such as lightning, tornadoes, and hurricanes.[26]

The vulnerability of nuclear power plants to sabotage or terrorist activities has been much disputed. Certainly the plants could be sabotaged, but so could fossil fuel plants. Unless nuclear plants were blown up so that radioactive materials were released into the atmosphere, sabotage would be no worse for nuclear plants than for fossil fuel plants.

Society is increasingly vulnerable to terrorist activities, and power plants—especially nuclear power plants—would furnish one possible target for militant dissident groups. Objectively, however, there is nothing especially or uniquely attractive about nuclear power plants for terrorist activities. It is not possible to steal any products from nuclear power plants from which a bomb can be easily fabricated. The wastes produced in a nuclear power plant contain plutonium from which a bomb could be constructed, but the actual wastes contain a mixture of a large number of radioactive substances, and plutonium could be separated from the other wastes only in a sophisticated and expensive processing plant. Thus, terrorists cannot use nuclear power plants as sources of fuel to build bombs.

In general, it can be said that the public has chosen to permit nuclear power and that as a consequence society will have to guard against terrorist activities. The decision of individual power companies whether to build fossil fuel plants or nuclear plants will depend in the main on the relative costs of the two types of plants. The possibility of terrorist activity may raise the cost of nuclear power plants, but it seems unlikely at current costs to tip the balance away from nuclear to fossil fuel plants.

Nevertheless, it can be safely predicted that any proposed nuclear power development will arouse the strong opposition of environmental and safety-oriented groups. Thus, power company executives may decide that the cost in time and effort of securing authorization from several regulatory authorities to construct a nuclear power plant is so great they would opt for fossil fuel plants, even if they are more expensive to operate. However, it should be noted that in general environmental groups strongly oppose any type of power plant, not just nuclear plants.[27]

In a nutshell, in most of the country nuclear power has an economic edge over fossil fuel plants. Unless political power is used to block atomic power plants or unless there is a major decline in the cost of fossil fuels, the great majority of future large power plants in most parts of the country, at least those constructed for base load, will be nuclear.[28]

---

[26] Ibid., p. 3.

[27] See, for example, G. C. Hill, "The Law's Delay: Plant's Demise Signals Troubles Ahead for Energy Expansion," *Wall Street Journal,* September 7, 1976, p. 1.

[28] Base load plants are those designed to operate near capacity constantly, as opposed to peaking plants designed to operate for only short periods of strong demand.

Nuclear Power's Share: Demand and Capacity Projections

Utility plans apparently are consistent with increases in capacity of about 5 percent per year—the rate forecast at the beginning of this chapter—and with a growing proportion of the new base load plants being nuclear. The Federal Power Commission collects from each of the utilities information on its projected major new plants. All planned fossil fuel plants larger than 500 megawatts capacity and all nuclear plants larger than 300 megawatts must be reported. Since it takes roughly ten years to build a new nuclear power plant and over six years for a fossil fuel plant (with the time becoming longer as there are more environmental and licensing hurdles to overcome), the future base load plants to be built by utilities can be reasonably well predicted.

Table 4 shows the expected rate of growth of major electric steam-generating plants of public utilities. Utilities will, of course, install smaller gas

Table 4

Projected Electric Generation Capacity
of Electric Utility Companies
(in megawatts)

Year	Total Capacity	Steam Capacity	Planned Steam Capacity by Utilities	Planned Addition to Capacity	Planned Nuclear Addition	Percent Nuclear Addition
	Realized	Realized				
1973	440	352				
1974	476	376	363	36.3	10.5	29.0
1975	505	396	389	26.0	7.3	28.0
	Projected[a]	Projected[a]				
1976	525	406	407	18.0	8.0	45.0
1977	551	426	427	20.0	7.0	35.0
1978	579	447	448	21.0	9.0	43.0
1979	608	470	469	21.0	8.0	39.0
1980	638	493	494	25.0	16.0	61.0
1981	670	518	522	28.0	20.0	72.0
1982	703	544	543	31.0	27.0	87.0
1983	738	571	564	21.0	18.0	88.0
1984	775	599	n.a.	—	n.a.	—
1985	814	629	n.a.	—	n.a.	—
1986	855	661	n.a.	—	n.a.	—

[a] Projected at a 5 percent annual growth rate.

Source: 1974 and 1975 data from *Statistical Yearbook of the Electric Utilities Industry for 1975*, Edison Electric Institute.

27

turbine and internal combustion plants for peaking purposes, but nuclear plants are not suitable for that type of activity. Thus, the second column shows the actual capacity of utility steam plants in 1973, 1974, and 1975 and projects the level of future capacity at a 5 percent growth rate. In the next column is the planned capacity of utilities during this period, assuming no major retirements of plants are carried out. Thus, the third column is computed by adding planned additions to capacity and previous years' capacity. As can be seen, future capacity as planned by the utilities is almost identical year by year with a projected growth rate of 5 percent. Since the planned additions to capacity were reported at the end of 1973, the capacity of fossil fuel plants planned for years after 1980 could be increased, but any nuclear power plants that are to come on stream by 1983 would have to have been planned by the end of 1973. Planned additions to capacity can be postponed if demand growth slows significantly, but it would be difficult if not impossible for utilities to increase capacity significantly above the levels in table 4 between now and 1980.

These figures indicate that for the next two years, 35 to 45 percent of all new capacity will come from nuclear plants. In 1980 and after, nuclear power will become even more important. These figures may overestimate dependence on nuclear power, since, as noted just above, fossil fuel plants that may come on stream in the mid-1980s would not necessarily have been planned by the end of 1973. Nevertheless, it seems safe to conclude that nuclear power will be an increasingly important element in future power production.

In 1974 the Atomic Energy Commission under various assumptions projected the growth in nuclear power capacity at the rate of around 18 percent.[29] During the same year they forecast that by 1990, 40 percent of U.S. generating capacity would be nuclear, up from about 6 percent at that time.[30] These estimates are much higher than others have predicted.

There is a wide variety of projections for both electricity demand and nuclear power plants. Most of them are based on past trends and forecast demand increases for electricity similar to the ones given above. However, one recent study deserves special notice, because it is more elaborate and sophisticated than most, and because it projects smaller increases in demand for electric power than usual.

Professors Paul L. Joskow and Martin L. Baughman, writing in the *Bell Journal of Economics* in 1976, reported on an elaborate model of the energy industries designed to forecast the demand for nuclear power.[31] They estimated the demand for electricity by regions and by customer classification (residential, commercial, and industrial). After forecasting capital costs and fuel costs,

[29] U.S. Atomic Energy Commission, *Nuclear Power Growth 1974-2000,* Report Wash-1139 (74), February 1974, p. 936.

[30] U.S. Atomic Energy Commission, *The Nuclear Industry 1974,* Wash-1174 (74), p. 14.

[31] Paul J. Joskow and Martin L. Baughman, "The Future of the U.S. Nuclear Energy Industry," *Bc `' Journal of Economics,* Spring 1976, pp. 3–32.

they estimated the impact these would have on prices and hence on demand for electricity. With the use of their computer simulation model, they projected electrical generating capacity, nuclear capacity, natural uranium utilization, and separative work requirements for eight cases. These cases included a base case with input prices at real 1975 values, a case with high air pollution restrictions, a case with peak load pricing enforced, a case with the lead time for nuclear plants decreased, a case of a nuclear moratorium, a case where the cost of uranium ore and enrichment rose sharply, and finally a case ignoring OPEC and leaving oil prices at the low levels existing in the 1970s. In all these cases, the necessary generating capacity and consequently the requisite nuclear capacity and the demand for separative work remained significantly below the AEC forecasts of 1974. Moreover, all of their estimates of capacity growth remained below the 5 percent level discussed above.

The Demand for Separative Work

The figures in table 4 can be converted into demand for separative work. Depending on the type of light water reactor, the original core of a 1,000-megawatt reactor would need from 365,000 to 434,000 kilograms of natural uranium enriched up to between 2 and 2.3 percent U_{235}.[32] At a tails assay of 0.25, this would take between 209,000 and 221,000 SWUs. The requirements for maintaining the reactor would be between 125 and 137 SWUs per megawatt. Assuming a plant factor (the percent of the actual power generated divided by the theoretical maximum during the year) of 0.7, the 1,000-megawatt reactor would use between 87,000 and 95,000 SWUs per year.

Of course, a higher tails assay would reduce the need for separative work while at the same time increasing the demand for natural uranium. The tails assay that will minimize costs will depend on the cost of enrichment and the cost of natural uranium, neither of which can be forecast accurately at this point. As pointed out in chapter 2, DOE is currently planning to use the new capacity from add-on facilities to keep the tails assay at 0.20 to 0.25 percent. The demand for separative work will also depend on whether uranium and plutonium are recycled from waste materials back into the generation of power. ERDA estimated in 1975 that plutonium recycling would reduce power generation costs by 14 percent.[33]

Assuming no plutonium recycling by the early 1980s, we can predict that demand by utilities in the United States will be around 16.3 million SWUs in 1983. On top of that will be a foreign demand in excess of 10 million SWUs, since DOE already has foreign orders (commitments) for 12.4 million in 1982,

---

[32] OECD Energy Agency and the International Atomic Energy Agency, *Uranium: Resources, Production, and Demand* (Brussels: OECD, 1975).

[33] House, Subcommittee on Energy Research, Development, and Demonstration of the Committee on Science and Technology, *Energy Facts II*, p. 259.

9 million in 1983, and 10.1 million in 1984.[34] Thus, it would appear that by the middle 1980s demand for enrichment services will conservatively exceed the projected capacity of the government plants.

Estimates have been made by several organizations. The OECD Nuclear Energy Agency and the International Atomic Energy Agency have estimated that the world demand for separative work could range from a high of 65 million SWUs in 1985 without plutonium recycling to a low of 51 million SWUs with recycling (see table 5 in chapter 4, below).[35] Most of the capacity to enrich uranium will have to come from the United States. They project that other noncommunist countries will have a total enriching capacity in 1985 of 27.2 million SWUs while the United States, in the absence of another plant, will have 27.7[36] Thus, if there is plutonium recycling—and it appears that European countries are moving ahead to recycle waste materials—and if demand grows at the slower rate, total world capacity will be enough in 1985 to meet demand but by 1987 world demand would equal or exceed world supply.

The 1977 report of the Nuclear Energy Policy Study Group also forecast both the world supply of SWUs and the world demand. They estimated that Western Europe's capacity to enrich uranium would grow much more slowly than table 5 suggests. Moreover, they forecast a much slower growth rate of demand and concluded that "it is very unlikely that there will be a shortage of enrichment capacity before 1988 and in fact unlikely that there will be one before the early 1990s if uranium is stripped to a tails assay of 0.3 percent or more uranium-235."[37]

ERDA estimated that the annual demand for separative work from domestic and foreign customers will exceed the projected capacity of the existing American government plants before 1985.[38] On the other hand, Joskow and Baughman's figures indicate that demand in the United States is going to grow more slowly than DOE estimates. Nevertheless, they conclude that "increased enrichment capacity will be needed before 1985 to meet foreign demands for enrichment services."[39]

Summary

Estimates of when additional U.S. enrichment capacity will be needed range from before 1985 (ERDA) to sometime in the early 1990s (NEPSG). ERDA

---

[34] House, Subcommittee on Energy of the Committee on Science and Astronautics, *Energy Facts,* p. 267.

[35] *Uranium Resources, Production, and Demand,* p. 69.

[36] Ibid., p. 66.

[37] *Nuclear Power Issues and Choices,* p. 369.

[38] ERDA, *Plea for a Competitive Nuclear Fuel Industry in the United States,* 1975, p. 2.

[39] Joskow and Baughman, "U.S. Nuclear Energy Industry," p. 25.

and its predecessor, the AEC, have consistently been overoptimistic about the growth of nuclear power, and it seems more likely that additional enrichment will not be needed before the late 1980s or early 1990s. Nevertheless, because of the long lead time involved in building new enrichment facilities, it is necessary to consider now how additional capacity will be made available, and by whom.

4

A Historical Perspective

The three government-owned uranium-enrichment plants were constructed for weapons purposes during World War II and in the immediate postwar period. Since then, there has been a gradual shift in the field of atomic energy toward the civilian private sector. In 1954 the Congress, through the Atomic Energy Act, established the policy of encouraging the use of nuclear energy for civilian peaceful purposes. During the 1950s and 1960s private enterprise took over more and more of the nuclear energy industry until only research and development and the uranium-enrichment process remained predominantly government responsibilities.

Under the Private Ownership of Special Nuclear Materials Act of 1964, the Atomic Energy Commission was authorized to offer enrichment services to the private sector. Enrichment technologies remained highly classified until 1967, when the commission, in order to assure the nuclear power industry of the adequacy of the gaseous diffusion plants, declassified information on the capacity of the plants. It subsequently released other technical information on enrichment.[1] In September 1967, the AEC announced that the toll charge for enrichment would be set at $26 per SWU. This charge was established on the basis of recovering certain government costs, including the direct cost of operating the plants, depreciation of that part of the plants directly used for enrichment for the private sector, process development, AEC administration, and imputed interest on investment and working capital.[2]

With virtually all of the nuclear fuel cycle in private hands except for the enrichment sector, the Atomic Industrial Forum, an industry group concerned with nuclear power, established in 1966 a Forum Study Committee to look into

---

[1] *AEC Gaseous Diffusion Plant Operations*, ORO-658, February 1968, p. 1.
[2] Ibid., p. 42.

the feasibility and desirability of transferring the gaseous diffusion plants to the private sector.[3] The committee concluded:

> The market for enrichment services . . . will be sufficiently broad so that there is a reasonable probability that a viable private industry can develop. In the opinion of the Study Committee, the fact that four or more separate enrichment plants will ultimately be needed creates an environment in which private enrichment services could be provided under competitive conditions. Competing privately-owned plants, under the spur of the profit motive, could be expected to make progress in cost reductions and to further the advancement of enrichment technology. In short, the Study Committee concludes that the declared Congressional policy in favor of private ownership and operation of facilities for the peaceful use of the atom can now be extended to uranium enrichment, the last remaining area of government ownership and operation.[4]

The AEC responded to this initiative by issuing its own staff report, *Future Ownership and Management of Uranium Enrichment Facilities in the United States*.[5] This report did not make any recommendation but outlined the national security considerations, the factors affecting the competitiveness of any enrichment industry, and the national economic and government financial interests.

In October of 1968, the chairman of the Joint Committee on Atomic Energy asked the General Accounting Office to evaluate the possible sale of the enrichment plants to the private sector. The report of the comptroller general, dated May 20, 1969, was generally negative toward the idea.[6] The report said:

> GAO's studies of the economic value of the gaseous diffusion plants indicated that . . . continued Government ownership of the three plants would result in the highest discounted net cash flow to the Treasury. . . .
>
> In GAO's opinion, the need for these very large financial commitments would limit the number of potential investors in the enrichment enterprise. . . .

---

[3]*Private Ownership and Operation of Uranium Enrichment Facilities,* Report of a Forum Study Committee, Atomic Industrial Forum, Inc., New York, N.Y., June 1968, p. IV., reprinted in *Selected Materials concerning Future Ownership of the AEC's Gaseous Diffusion Plants,* Joint Committee on Atomic Energy, 91st Congress, 1st session, Joint Committee Print, June 1969, pp. 379-456.

[4]Ibid., pp. 2-3.

[5]*Summary Report by AEC Staff on Future Ownership and Management of Uranium Enrichment Facilities in the United States,* March 1969, reprinted in *Selected Materials concerning Future Ownership of the AEC's Gaseous Diffusion Plants,* pp. 1-32.

[6]*Report to the Joint Committee on Atomic Energy, Congress of the United States, Possible Transfer of the Atomic Energy Commission's Gaseous Diffusion Plants to Private Ownership by the Comptroller General of the United States, May 20, 1969,* reprinted in *Selected Materials concerning Future Ownership of the AEC's Gaseous Diffusion Plants,* pp. 271-374.

GAO believes that, from the standpoint of ensuring that additional enrichment plant capacity will be available when needed, early transfer of the three existing plants to private ownership would be a less favorable approach than continued Government ownership. However, GAO noted that, under AEC ownership, delays had been encountered in implementing the CIP [Cascade Improvement Program], in part because funds for the program must be obtained through the budgetary process.[7]

In the spring of 1969, the White House appointed a task force headed by the chairman of the Council of Economic Advisers, Paul W. McCracken, and including the director of the Bureau of the Budget, the director of the Office of Science and Technology, a commissioner of the Atomic Energy Commission, and representatives from the Department of State, the Antitrust Division of the Justice Department, and the Treasury.[8] The report of this task force, dated August 1969, was initially classified, but at the request of Representative Chet Holifield, chairman of the Joint Committee on Atomic Energy, an unclassified version was made available to the Joint Committee. The task force did not reach any conclusions but attempted to analyze the issues, identify the primary alternative courses of action, and present the advantages and disadvantages of each. It stated:

> Early sale [of the enrichment facilities] to private industry would involve establishing initially either a monopoly or a three-firm oligopoly. If the oligopoly were established initially the éntry or serious threat of entry of new firms might be possible in the late 1970's or early 1980's but until that happened the industry would not likely be sufficiently competitive to keep prices in line with average costs. . . .
>
> Maintenance of the present system would continue the practice of charging rates designed to recover government costs but would lead to prices below total economic costs to society. In addition, under a continuation of the present system there is less commercial incentive to keep costs down and to do R & D, although national security incentives to do R & D will continue. Moreover, if the financing of capacity expansion continues to be dependent upon the appropriation process, capacity may not be expanded in time to meet the forecasted growth in the most economical manner.[9]

Notwithstanding a certain sympathy within the Nixon administration for the transfer of the plants to the private sector, the strong congressional opposition to the sale of the plants led to a decision to keep the plants in government

---

[7]Ibid., pp. 1-3.

[8]The author was the CEA staff representative and the principal author of the report.

[9]*Report of the Task Force on Uranium Enrichment Facilities,* unclassified version, Council of Economic Advisers, June 29, 1970, pp. 2-3.

hands. An attempt by the administration and the AEC was made to put the charges for enrichment services on a more economic basis by changing the pricing formula to include an estimate for the cost of new facilities, taxes that would be paid in the private sector, and private market interest rates of an appropriate risk. Congress quickly passed an amendment to an appropriation bill returning to the previous pricing system.[10]

In February of 1972, James Schlesinger, then chairman of the Atomic Energy Commission, announced that it was time for the private sector to enter the enrichment business to meet future demand. The AEC subsequently established two programs to encourage the private sector to develop the expertise and knowledge necessary to build enrichment facilities. The Joint Committee on Atomic Energy held hearings on the development of the future structure of the uranium-enrichment industry in 1973, 1974, 1975, and 1976. Thus, the stage was set when Uranium Enrichment Associates (UEA), a consortium of Goodyear Tire and Rubber Company, Williams Company, and Bechtel Corporation (which took the lead), proposed a privately built gaseous diffusion plant. Before undertaking construction of a mammoth enrichment plant at a cost of $3.5 billion, however, UEA wanted significant guarantees.

The Ford administration responded by proposing that the government be authorized to enter into cooperative arrangements with private firms that wish to build, own, and operate uranium-enrichment plants. The proposed Nuclear Fuel Assurance Act (NFAA) would have permitted the government to supply and warrant enrichment technology; to sell materials available from the government on a full cost basis; to buy enriching services from private producers or to sell enriching services to producers from the government's stockpile in order to accommodate plant start-up and loading problems; and to guarantee the delivery of uranium-enrichment services to customers who had placed orders with private enrichment firms. Moreover, if a private enrichment project was in danger of failure, the government, either at the request of the private enricher or on its own initiative, could have assumed the assets and liabilities of the venture, with compensation to the domestic investors depending on the circumstances leading to the threat of failure. Any contract signed with a proposed private enricher would have been subject to congressional review and approval before becoming effective. The provision permitting the government to acquire the assets and liabilities of a private venture either at its own initiative or at the initiative of the private company would have come to an end after a period sufficient to demonstrate commercial operation.

The UEA proposal that stimulated this administration response consisted of considerably more than simply opening the market to the private sector. Basically, UEA wanted a number of very significant, subsidylike provisions designed to minimize its risk.

---

[10]*Wall Street Journal,* February 18, 1977, p. 14, col. 3.

The UEA plan was to construct a gaseous diffusion plant with a capacity of 9 million separative work units near Dotham, Alabama.[11] This plant would have required about 2,500 megawatts of electrical power, which would have been supplied from two nuclear power facilities to be constructed nearby for that specific purpose. The enrichment plant, when in full operation, would have provided service to about ninety large nuclear power reactors. UEA estimated in 1975 that the cost, exclusive of the electric power project, would be $3.5 billion in 1976 dollars. The schedule called for construction to start in 1977, limited production to commence in 1981, and full production to be reached by 1983. It was expected that about 60 percent of the plant's output would be sold abroad.

To finance the construction of this plant, UEA planned on a debt-equity ratio of 85:15, with foreign buyers supplying 60 percent of each. Even though 60 percent of the equity was to be held by foreign participants, they would have held only 45 percent of the voting rights. The major foreign countries that expressed interest and the maximum potential financial contributions were: Iran (20 percent), Japan (20 percent), West Germany (11 percent), and France (10 percent).

UEA also expected to negotiate with utilities a pricing formula that would have recovered all operating costs, serviced the debt, and provided a return of approximately 15 percent on equity, after taxes. The government would have received a 3 percent of gross revenue payment for the use of the taxpayer-developed technology. UEA also planned to organize a wholly domestically owned corporation, Uranium Enrichment Technology, which would have access to and exercise control over classified enrichment technology.

The executives of UEA claimed that financing could not be arranged for their uranium-enrichment plant without government assurances because of the long period before investors could expect any return on capital, the great uncertainties connected with a new technology and a new industry, and the possibility that regulatory action might curtail the final market. Thus their original proposal, which was significantly modified before the plan was abandoned in the fall of 1976, would have had the government supply, at cost, essential components produced exclusively by the government, such as enrichment barriers and seals. The government would also have supplied gaseous diffusion technology and warranted its satisfactory operation. Moreover, the government was to be prepared to purchase enriching services from UEA if private demand was inadequate at the time the plant came on line and to sell enriching services to UEA from the government's stockpile to accommodate

---

[11] The description of the UEA plan is taken from the White House ''Fact Sheet'': The President's Plan for a Competitive Nuclear Fuel Industry,'' issued June 26, 1975, pp. 16-19. A description can also be found in Comptroller General of the United States, *Evaluation of the Administration's Proposal for Government Assistance to Private Uranium Enrichment Groups,* report to Joint Committee on Atomic Energy, U.S. Congress, RCD-73-36, October 31, 1975.

plant start-up and loading problems. The latter provision, UEA argued, was necessary in order to assure customers that if construction of the gaseous diffusion plant was delayed they would still be able to secure enrichment services.

In addition to these assurances, UEA wanted the government to be obliged to purchase the domestic owners' controlling interest at the consortium's request. Furthermore, if it were in the national interest the government could have taken over ownership at its own initiative. These two options would have expired one year after full commercial operation was demonstrated. Finally, if UEA should have been unable to complete the plant or to bring it into commercial operation, the government would have provided the funds necessary to finish it. If this failure was not UEA's fault, the government would have had to take over the domestic equity and assume all of UEA's liabilities and debts; if the government's acquisition of the plant was due to uncorrected gross mismanagement by UEA, no compensation for equity would have been necessary.

Reactions to the Bill

There was considerable controversy over the Nuclear Fuel Assurance Act. While some worried about the potential risk that secret technologies might be revealed to other nations, the main objection dealt with the extensive guarantees to be offered private industry. Senator John Pastore, chairman of the Joint Committee on Atomic Energy, asked the comptroller general of the United States, Elmer B. Staats, to evaluate the administration's proposal. The final General Accounting Office (GAO) report was quite negative and was especially critical of the risk shifting from private industry to the government.[12]

The comptroller's report pointed out that the agreement proposed between ERDA and UEA would result in a potential financial commitment to reimburse domestic participants if UEA could not finish the project and to purchase up to 6 million SWUs from UEA. Furthermore, in the admittedly unlikely event that the government took over the project and it proved unsalvageable, the Treasury would have had to cover the cost of the two nuclear power plants to be dedicated to the enrichment plant, less any revenues that could be garnered from the sale of electricity.

The risk for domestic investors in the project would be comparatively small: to have lost all their investment would have required gross mismanagement, gross negligence, or willful misconduct. For a finding of any of these, the government would have had to send a formally written notice of deficiencies to UEA, and UEA must have failed to respond reasonably to the notice. A partial loss of equity could have occurred, depending on UEA's compliance with its commitments, the efforts of UEA, and the degree of fault.

---

[12]Comptroller General, *Evaluation of the Administration's Proposal.*

On the other hand foreign participants would have had much more at risk than would domestic equity and debt holders. The U.S. government would not have taken over their equity or debt even in the event that the domestic investors were being reimbursed. Moreover, if substantial cost overruns occurred, foreign countries would have been expected to continue to provide their prorated share of the funds to complete the plant.

The reader may wonder why foreign countries would have agreed to participate in such a venture, especially where domestic investors were almost completely protected. Apparently UEA believed that the projected worldwide shortage of enrichment capacity would have forced foreigners to invest in order to secure enrichment services for their nuclear plants. If worldwide demand for enrichment services grows as expected and other countries fail to revise their plans to expand their enrichment capacity significantly, UEA's expectation may prove true by the late 1980s. Jerome W. Komes of Bechtel Corporation, who was one of the major spokesmen for UEA, claimed in April of 1976 that two-thirds of the 60 percent share that foreigners were expected to invest had already been agreed upon.[13]

Table 5 shows the OECD's estimates of worldwide demand for enrichment assuming both plutonium recycling and no plutonium recycling, with high and low estimates. It also shows the expected enrichment capacity, not including a new American plant or any ERDA-built add-on facility. Figures for the two major European enrichment combines, URENCO and EURODIF, are presented separately.

URENCO, as noted in chapter 2, is a combine of the Netherlands, West Germany, and the United Kingdom formed to build and operate gas centrifuge enrichment plants. Substantial prototype plants have been operated both in the Netherlands and in Great Britain since 1972 and full production is expected by 1978. The expansion potential is limited only by financial and market considerations, with construction lead times of four to five years before plants are fully on line.

The EURODIF operation is a joint French, Italian, Belgian, and Spanish gaseous diffusion plant under construction in France with an expected full capacity of 10.8 million SWUs. It is expected to be on line by 1981, and a second plant is under consideration but with the site, partners, and organization still undecided.

As table 5 shows, if there is no plutonium recycling and if the demand grows at the high estimate, by 1982 world annual demand will exceed world capacity. Since the U.S. government is building a stockpile of enriched uranium that can be drawn on, there need be no immediate shortage in that year, but the shortfall will continue to grow and cannot be long met from the stockpile. On the other hand, if plutonium is recycled and if demand grows at

---

[13] Interview with author, April 12, 1976.

Table 5
World Separative Work Supply and Demand
(millions of SWUs)

| Year | Annual Requirements[a] | | | | Annual Capacity | | | | |
	Without Pu Recycle		With Pu Recycle		Existing ERDA Plants	URENCO	EURODIF I & II	Other	Total
	High	Low	High	Low					
1976	12	12	12	12	20.3	0.1	—	0.4	20.8
1977	14	13	14	13	22.5	0.2	—	0.4	23.1
1978	18	16	18	16	24.6	0.4	—	0.4	25.9
1979	23	21	23	21	26.5	0.7	3.1	0.4	30.7
1980	31	28	31	28	27.7	1.0	6.5	0.4	35.6
1981	38	35	35	32	27.7	1.5	8.4	0.4	38.0
1982	44	39	41	37	27.7	2.5	10.8	0.4	41.4
1983	52	46	48	43	27.7	4.0	10.8	0.4	46.4
1984	58	52	53	47	27.7	7.0	13.8	0.4	49.4
1985	65	57	58	51	27.7	10.0	16.8	0.4	54.9
1986	74	65	66	58	27.7	10.0[b]	19.8	5.4	62.9
1987	82	71	73	63	27.7	10.0	19.8	5.4	62.9
1988	92	80	82	71	27.7	10.0	19.8	5.4	62.9

[a] Assuming 0.25 percent tails assay and 70 percent local factor.
[b] Capacity will be increased according to requirements.

Source: OECD Energy Agency and the International Atomic Energy Agency, *Uranium: Resources, Production, and Demand*, tables 14 and 21.

40

the slowest rate, annual demand will not exceed world capacity prior to 1987.

As can be seen from this table, if the UEA plant were built according to the original schedule and reached full production of 9 million SWUs in 1983, there is an excellent chance that there would be no need for its output, but by the mid-1980s there would be. To cover this contingency, as noted earlier, the UEA proposal required that ERDA be prepared to purchase up to 6 million SWUs during the first five years of UEA's plant operation. This would have been a potential commitment of $1.2 billion by the American government.[14] But UEA planned to negotiate binding contracts with utilities and foreign governments to purchase separative work, and it would have been unlikely for the company to start construction of the facilities before most of its output had been sold.

Domestic utilities would probably have been reluctant to enter into such binding contracts with UEA, since the price for enrichment services was to be based on a cost-plus contract. As already noted, UEA had planned to contract with its customers to charge them a price for enrichment services that reflected a 15 percent return on equity after all federal, state, and local taxes had been paid. Since equity was to have been equal to 15 percent of the total cost of the project, the larger the equity, the larger would be the profit earned by the consortium (provided the cost of equity capital did not exceed 15 percent). Higher capital costs would have increased equity and thus profits. Higher operating costs would not reduce profits but would simply lead to higher prices for enrichment services. Moreover, UEA expected to negotiate contracts with customers that would have included a take-or-pay provision, thus assuring UEA of the sale of all the contracted enrichment services, no matter what its price and irrespective of the price of alternative supplies of enrichment services. Of course, utilities would probably have negotiated some control over project costs.

It is hard to predict what would have happened had the UEA project gone forward. If utilities and foreign investors believed that enrichment services would be available only through UEA, they would have swallowed the harsh terms. But if other sources of enrichment looked sufficiently likely, they would have been unwilling to enter the contracts UEA wanted, and the gaseous diffusion plant might never have been built. In hindsight it does not appear that utilities would have had to rush into agreeing to UEA's terms. They might instead have looked to some of the proposed centrifuge enrichers for services.

Other Effects of the Bill

While the Nuclear Fuel Assurance Act was tailored to accommodate the UEA proposal, it was also intended to encourage other enrichment projects. Three other firms and consortia, Centar, Exxon Nuclear, and Garrett Corporation, submitted proposals for cooperative arrangements with the federal government

[14]Ibid., p. 17.

to build gas centrifuge plants. As explained in chapter 2, while gas centrifuge technology has been not fully demonstrated on a production basis, it is expected to be the preferred technology for future additions to enrichment technology. In fact, the Carter administration has announced that the next addition to the capacity will be centrifuge.

None of these three proposals was as well developed as UEA's, so less is known about the type of government guarantees that would have been needed. The NFAA did, however, envision contracting with these or other firms for the construction and operation of gas centrifuge plants. ERDA asked organizations that desired to construct uranium-enrichment plants to submit proposals by October 1, 1975. Proposals were received from Garrett Corporation, a manufacturer of equipment to generate, transform, and control energy; Centar Associates, a joint venture of Electro-Nucleonics and Atlantic Richfield Company; and Exxon Nuclear Company, a wholly owned subsidiary of Exxon Corporation.

Garrett has participated in research and development of uranium enrichment under government contract since 1961. It has completed a pilot centrifuge machine manufacturing line and is supporting the pilot centrifuge enrichment plant at Oak Ridge.[15] Its wholly owned subsidiary, Texas Regional Enrichment Corporation, submitted a proposal to build a 3-million-SWU centrifuge plant in Texas that would initially have produced about 350,000 SWUs in mid-1981 and would have expanded its capacity by 1987. Garrett officials told General Accounting Office investigators that they would request government "assurance in the areas of (1) process procedures, (2) completion guarantees, and (3) some early access to the government SWU stockpile."[16]

Centar, one of whose parent corporations has been in the centrifuge business since 1963, proposed to build a 3-million-SWU centrifuge plant at a cost of about $1 billion which would come fully on line by 1986. In the first stage, about $100 million would be invested in a prototype plant to test the economics of gas centrifuge technology.[17] Centar wanted the government to temporarily underwrite the debt portion of the financing in the form of guaranteeing the government's technology. The company told GAO that it was prepared to accept the loss of its 25 percent equity investment in the project if it failed, except in the case of a government action which precluded "Centar's continuance as a commercial venture."[18] A supply of SWUs from ERDA's stockpile would have been needed to support early production contracts.

Like the others, Exxon Nuclear Company proposed to construct a $1 billion, 3-million-SWU centrifuge plant with initial operation of about 1 mil-

---

[15]Comptroller General, *Evaluation of the Administration's Proposal,* p. 23.

[16]Ibid.

[17]*Nuclear Energy,* June 1975, p. 13.

[18]Comptroller General, *Evaluation of the Administration's Proposal,* p. 23.

lion SWUs aimed for the 1981-1982 period. Full production would have been reached several years later. Exxon also required government guarantees for the process, an appropriate climate for freely buying and selling SWUs on a commercial basis, completion guarantees, and a government promise to pick up defaulting utility obligations.

As far as can be judged at this point, these centrifuge proposals did not require as extensive government guarantees as UEA was requesting. The companies were apparently not asking that the government take over their equity if their project failed economically. The main guarantees they wanted seemed to involve assurances that the processes would work as the government claimed. Since the proposals were never as well developed as UEA's, however, we cannot be sure what they would have required.

Both Presidents Nixon and Ford, as we have seen, made efforts to move the enrichment policy into the private sector. Both presidents failed because private firms were not eager to take up the challenge and did not make a concerted effort to change the status quo. Until business seriously supports such a change, none is likely to take place.

5

Uranium-Enrichment Options

The government faces four basic options for the future of uranium enrichment. It may continue its present policy, with control vested in a government agency such as the Department of Energy (DOE). It may establish a government corporation that, while wholly government owned, would be run on a profit-making economic basis. It may continue to operate the existing plants and the planned add-on facility while permitting and encouraging private enterprise to build new capacity that will be needed by the late 1980s. Finally, it may sell the existing plants to the private sector and get out of the enrichment business altogether. This chapter will analyze each of these alternative policies and weigh the costs and benefits of each.

The Carter Administration Plan

The Carter administration appears to have opted for continued government ownership and operation of the enrichment business. As noted earlier, it has decided that the Department of Energy will construct a centrifuge plant with four 2.2-million-SWU modules as an addition to the Portsmouth facility.[1] The administration and Congress are, however, considering having two to four private contractors operate the centrifuge units, with the idea that eventually the private sector could own enrichment facilities.

The chief advantage of continuing the present government policy is that it is the easiest and cheapest option from the point of view of government officials. No new legislation outside of annual appropriation bills will be necessary. Existing skilled and knowledgeable personnel will continue to run and plan the enrichment operation. Not only can new facilities be built to satisfy the existing demand, but uneconomic facilities can be built to try to forestall foreign enrichment. From the security perspective, the fewest number of

---

[1]*Nuclear Industry,* August 1977, p. 14.

individuals will become knowledgeable about the few remaining secret processes involved in enrichment, though it should be noted that there is general agreement that security issues are not really important in this area.[2] Also, even if full "privatization" were authorized, controls could still be maintained on those processes where security is important to the national interest.

The nuclear power industry would prefer to maintain the status quo, as they foresee that separative work would be more expensive and perhaps less available in a private market.[3] In addition, the reactor manufacturers would for the same reason prefer to keep the enrichment business in government hands and under the control of the government.[4] Thus influential vested interest groups will oppose any change.

There are, however, a number of significant drawbacks to continuing uranium enrichment under DOE's auspices. In the past, Congress and the executive branch have been supportive of nuclear energy. The current plan for the construction of additional enrichment facilities at Portsmouth, Ohio, at an expected cost of between \$4.2 and \$4.5 billion,[5] is evidence of their continuing support. Nevertheless, other additions to enrichment facilities, which will be needed by the late 1980s, may be harder to come by. Reflecting the view of the Joint Committee on Atomic Energy, Congress has in the past strongly supported nuclear power, but the Joint Committee was abolished in early 1977 and its power distributed to less sympathetic congressional committees. Moreover, environmentalists have become increasingly influential. Small but vocal groups are adamantly opposed to nuclear power and will attempt to block the construction of any facilities that might enlarge the industry.[6] Even though the initiatives placed on the ballot in six states in 1976 were overwhelmingly defeated, these groups have vowed to continue the fight.[7]

While in the end Congress will probably be willing to construct additional enrichment facilities, the victory will come only after a battle that may result in a real slowdown of construction. One of the main factors that has led the nuclear power industry in the past to support continued government operation and construction of enrichment facilities was the belief that more separative work at a lower price would be available under government ownership than if the private market provided the service. No doubt they were right, but with the new political force of the antinuclear establishment that may no longer be true.

---

[2] Comptroller General, *Evaluation of the Administration's Proposal,* p. 30.

[3] Statement by Gordon R. Corey of Commonwealth Edison Co., in U.S. Congress, Joint Committee on Atomic Energy, *Nuclear Fuel Assurance Act,* Hearings before the JCAE; (Washington, D.C.: U.S. Government Printing Office, 1976) pt. 2, pp. 28-30; also letter from R. C. Seamans, ERDA, to Chairman Senator J. Pastore in ibid., p. 450.

[4] Ibid.

[5] *Nuclear Industry,* August 1977, p. 14.

[6] An example is the opposition to the Jamesport nuclear project by the Long Island Lighting Co., reported in *New York Times,* September 5, 1976, p. 40, col. 3.

[7] *New York Times,* June 10, 1976, p. 55, col. 3.

The antinuclear lobby would also affect a private enrichment industry but to a lesser extent. The Congress is more subject to lobbying pressure than an independent body such as the Nuclear Regulatory Commission. A private enrichment industry would need only NRC approval to expand.

The political argument over the construction of nuclear plants and enrichment demonstrates one of the more serious drawbacks to government ownership of commercial enterprises: decisions tend to be made on political grounds rather than on the basis of economic considerations. Political decision making can lead to unneeded facilities, the construction of facilities at uneconomic locations, or even the blocking of economically justified plants. Political considerations are quite common in other government-run enterprises. For example, the postal service has been subject to intense pressure to maintain uneconomic rural post offices;[8] Amtrak has been forced to maintain uneconomic service to communities represented by influential congressmen;[9] the military has had considerable difficulty in closing uneconomic naval yards, arms facilities, and military bases.[10]

Not only is the construction of new uranium-enrichment facilities subject to political pressure but the pricing of the separative work has been subject to congressional pressure and would continue to be. In 1970 the Atomic Energy Commission announced a new pricing formula based on economic considerations.[11] The Joint Atomic Energy Commission, believing that the provision would lead to significantly higher charges for enrichment, wrote into an appropriation bill an amendment barring the new pricing formula and mandating the old.[12] This is a particularly acute problem in cases where a government agency runs the facilities. Part of the objective of setting up Amtrak and the postal service was to provide them with more pricing flexibility than they would have as a branch of the government, but this does not seem to have made much difference.

The Carter administration has recently proposed raising the price of separative work from $61.30 to $100 per SWU, although Congress may veto the increase. ERDA says that its "fair value pricing" is based on a formula that includes costs of taxes and insurance, interest rates at a cost private companies would have to pay, and the cost of a new plant.[13] This formula, if put into effect, would reflect most of the factors that go into a market price but no government agency, no matter how willing, can ever duplicate the subtleties of a private market. Besides profit, which is ignored in the formula, the policy does not give

---

[8]Ibid., September 24, 1976, p. 41.

[9]*Wall Street Journal,* December 28, 1976, p. 6.

[10]Witness the House of Representatives vote to halt Defense Department plans to cut back or close thirteen military bases, described in *New York Times,* May 8, 1976, p. 9, col. 1.

[11]*Wall Street Journal,* December 24, 1970, p. 6, col. 1.

[12]Ibid., February 18, 1971, p. 14, col. 3.

[13]*Nuclear Industry,* August 1977, p. 14.

guidance to the most appropriate contract terms which can be as important as the nominal price. It should be stressed that profit is a proper return on equity that reflects the inherent risk in an enterprise; any formula that does not include profit will result in an underestimate of the competitive market price.

Another major drawback of government operation is that, once a project is started, individuals and groups develop vested interests in it and oppose vigorously any attempt to cancel or significantly modify it, even in light of overwhelming evidence of the wastefulness of the operation. For example, the British have been unwilling to cancel the wasteful and extremely expensive Concorde project, even knowing that each plane produced adds to their losses.[14] In the United States, study after study has shown that fixed rail transit systems such as BART in the San Francisco Bay area and the Washington Metro System are wasteful and costly failures by any criterion,[15] but they continue to be built and operated.

The waste inherent in government programs is exemplified in the work by ERDA on the liquid metal fast breeder reactor (LMFBR). A recent study concluded:

> Many of the AEC's projections lead to an unrealistically large benefit from the LMFBR: uranium resources are underestimated. The high-temperature gas reactor is artificially restricted to a low level of participation in the future electric power system. Future energy demand is overestimated. The plant capital cost of the LMFBR is decreased too rapidly to fit any reasonable learning curve. And finally, schedule slippages and cost overruns are not adequately reflected in the analyses.
>
> It thus appears that the LMFBR's high efficiency in uranium utilization is not sufficient to compensate for its higher plant capital and program costs. The LMFBR program yields no net discounted economic benefits.[16]

Government managers have little incentive to minimize cost or to innovate. Cost decreases reduce the budget allocation that will be authorized by the Office of Management and Budget and by the Congress. Such decreases for a single project will have little impact on the overall government budget but will no doubt disturb some existing program or officials. Thus, putting such cost reductions into effect will be costly to the government official, will have little overall impact, and will result in a reduction of his or her budget which is likely to be viewed as undesirable by the bureaucrat. Innovations and new procedures also are unlikely to be used or implemented because of the relatively high cost to

[14]*New York Times,* February 1, 1973, p. 53, col. 5.

[15]"BART," *New York Times,* October 19, 1975, sec. 6, p. 17.

[16]Brian G. Chow, *The Liquid Metal Fast Breeder Reactor: An Economic Analysis* (Washington, D.C.: American Enterprise Institute, 1975), pp. 71-72.

the official in charge and the relatively small return to him or her reducing costs or making innovations.

Moreover, in cases where the plants are operated by private companies on a fixed fee, there is no incentive for the operators to reduce costs, manpower, or to innovate.

It is for these reasons that most observers have viewed government operation of commercial activity as inefficient. If for overriding reasons the government should decide to own an enterprise, the Congress and most observers have felt that a separate government corporation providing some insulation from politicization of decision making (however slight) would be the preferred form.

A Government Corporation

Government corporations are less vulnerable to political pressure than operations carried out by departments which must request annual congressional appropriations. Nevertheless, pressure can and is often brought to bear to fulfill some special "social" need. At times, special subsidies are appropriated for the purpose of providing uneconomic services (for example, the postal service receives a subsidy to cover the carriage of some mail at reduced rates). At other times, Congress has coerced a corporation to subsidize specific services from its receipts on other services (for example, Amtrak has been pressured into providing little-used passenger service to West Virginia).[17]

Nevertheless, the objective of establishing a government corporation is to put the operation on a commercial basis that will result in its earning enough to cover its costs. Unlike a true commercial enterprise, profit maximization is never a goal, but the pressure to cover costs and to be independent of the appropriation process does provide some stimulus to reduce costs and to market the service efficiently.

Thus, a government corporation provides some incentives for using resources efficiently and for carrying out research and development programs, provided it does not have a monopoly,[18] but the incentives are weaker than in the private sector and can be diverted by acts of Congress. In this case the government enterprise would have a monopoly, which would further weaken any incentives to carry on research and development work. Moreover, even if it does attempt to minimize costs, the costs it minimizes are not the costs to society. A private firm must pay its labor and capital at least what they could earn under alternative conditions. Thus, a private firm, in minimizing expendi-

---

[17] *New York Times*, December 1, 1974, sec. 4, p. 9, col. 1.

[18] Douglas Adie, *An Evaluation of Postal Service Wage Rates* (Washington, D.C.: American Enterprise Institute, 1977) shows that monopoly power with no profit incentive for managers in a government corporation prevents both efficiency and willingness to prevent monopoly pricing.

tures, minimizes the cost of its output to the economy. The fact that the prices it pays equal or exceed what the resources could earn in the next best alternative implies that in minimizing its costs it utilizes the resources in a manner that economizes on them.

But when the government borrows—even a government corporation— lenders do not view their loans as risky. The government can always pay its debts by printing more money. Thus the interest a government corporation pays does not reflect the inherent risk in the government enterprise. Capital is therefore diverted from other, perhaps less risky, operations to the government enterprise. The managers of the government corporation, responding to the low cost of capital, will tend to invest more heavily in labor-saving and capital-intensive machinery and in monumental construction than is optimal. The low interest rate can be used to "justify" investment in uneconomic projects. As a consequence, even the best-run government corporation, managed without interference from either the executive or the legislature, will not use resources efficiently. Too much capital will be used in any project,[19] and some wasteful projects will be undertaken.

Another major drawback in any government operation, including a government corporation, is the problem of executive and congressional oversight. Sensitive areas such as pricing, labor policies, and construction of new facilities are usually subject to congressional hearings and sometimes to budget review. For example, in July 1977 the Senate voted down (58 to 39) a proposal to ban President Carter's proposed commercial pricing formula but passed an amendment that would give either house sixty days to veto any increase.[20]

Purchasers of a government-provided service bring political pressure to bear to maintain low and favorable rates. Labor unions clamor for higher wages, better working conditions, and more job security. Thus, it is more difficult if not impossible for the most economy-minded and commercially oriented managers to do an economical job of running the government enterprises.

"Privatization" of New Capacity

A mixed approach to enrichment is another policy option. The Ford administration's proposed Nuclear Fuel Assurance Act was an example of this type of proposal. In a mixed approach, the government would continue to operate the existing plants and perhaps the Portsmouth add-on facility, while additional new capacity would be constructed by the private sector. The Ford proposal was discussed in chapter 4 and therefore need not be gone into in detail here.

It is apparent, however, from the experience with the Ford proposal that

---

[19]Richard Muth, *Public Housing* (Washington, D.C.: American Enterprise Institute, 1973).
[20]*Nuclear Industry,* August 1977, p. 14.

private investors would require guarantees that the potential market for enrichment services would not be negated by government SWU regulatory actions. In addition, since there is no experience with enrichment as a commercial, profit-making enterprise, the private sector has asserted a strong need for government assurances of the viability of the processes and of the government acting as a market for enrichment services—buying separative work if supply is temporarily in excess and providing enrichment if demand is temporarily in excess.

Given the political opposition that developed in 1969 when the possibility of selling the government enrichment facilities was being studied, it is obviously less costly politically to let private enterprise build additional facilities while the government keeps the existing plants. In such a program, there can be few claims that the private sector will benefit from a government "give away." In addition, by keeping some facilities in government hands, those worried about the possible need for highly enriched uranium for military purposes can be reassured.

Moreover, permitting the private sector to build new capacity would mean that commercial economic principles would start being applied to the enrichment business. The private sector would have an incentive to minimize costs and to innovate. To the extent that competition developed, prices would be held down and marketing would be efficient.

Over time, the market could be expected to grow competitive, but in the beginning of "privatization" there might be more of a problem. If the Carter administration promptly moves to authorize private enrichment plants, new private enrichment facilities could be in operation by 1985, which is approximately when annual demand will begin to exceed annual capacity, judging from table 4. Under the Ford administration proposal, four companies or consortia proposed to build enrichment facilities. If the construction schedule proposed by these private companies were followed with a lag of about a year or two, UEA would be the only private enricher for about twelve months, but Centar, Garrett, and Exxon would not be far behind. Recently Exxon Nuclear Corporation applied for authorization to construct an experimental laser facility at Richland, Washington. This plant would have a capacity of 20,000 SWUs annually with a limited start-up in late 1980. If the facility resolves certain questions, design work on a 2-million-SWU capacity plant could be started.[21]

There are a number of other entities that may construct enrichment plants, and the firms already discussed above may also expand their planned capacities.[22] Uranium Enrichment Corporation of South Africa (UCOR) has a unique secret enrichment process and plans a 5-million-SWU plant that will be opera-

---

[21]Ibid., September 1977, p. 19.

[22]Information on other potential enrichers is drawn from Comptroller General, *Evaluation of the Administration's Proposal*, appendix 1, pp. 48-49, and from OECD Energy Agency and the International Atomic Energy Agency, *Uranium: Resources, Production, and Demand*, pp. 62-63.

tional by 1988. Australia, a potential major supplier of uranium, plans to construct an enrichment plant, probably centrifuge, to process its own uranium. In Canada two groups are considering constructing gaseous diffusion plants to come on stream in the mid-1980s. The Japanese have been doing research on centrifuge technology with the aim of starting a commercial separative work plant by 1985 with a capacity of 1 million SWUs. In fact, in November of 1976 three Japanese firms tentatively agreed to develop jointly centrifuge separators to be used in Japan's first enrichment plant.[23] West Germany and Brazil are planning enrichment plants using the jet nozzle technology. A West German firm first announced plans in November to construct a centrifuge enrichment plant with a capacity of 1 million SWUs.[24] Finally, it should be noted that the largest enrichment capacity outside the United States is in Russia. It has been estimated that their annual capacity is about 7 or 8 million SWUs. Most of their capacity is used behind the Iron Curtain, but they have made a number of sales in the West.

As can be seen, there are a large number of potential enrichers. While not all of them will actually build plants, by the late 1980s there could be a substantial number of private and public firms in the enrichment business. From the point of view of sheer numbers, there would be enough firms to provide a highly competitive industry. Shipping costs of enriched uranium are low compared to the value of the material. Unless artificial barriers are erected to foreign enrichment, the market for enrichment services should be worldwide.

Any discussion of the potential competitiveness of this industry must consider the fact that most enrichment services will be sold on long-term contracts. However, at the time the contracts are negotiated purchasers can look to numerous suppliers, even though none may actually be in business. In the next few years, as potential suppliers go ahead with their plans to construct enrichment facilities, there should be considerable competition to line up utility customers. No plant will be built without already having sold most of its capacity for the first twenty-five years.

Unless the demand for nuclear power grows even faster than the OECD's high estimate, it will not be possible for all of the projected enrichment projects to go forward and come on stream in the last half of the 1980s. Actually it would seem more likely that the growth in nuclear power will be at the low end of their estimate. Furthermore, there is a strong worldwide opposition to nuclear power by influential groups. While the defeat of the various state initiatives indicates that these groups cannot stop nuclear energy, they can slow its construction and make fossil fuel plants more attractive to utility managers. As shown in chapter 3, nuclear power is to a very real degree competitive with fossil fuel plants. Thus, demand for separative work is by no means completely inelastic, and it would seem highly likely that there will be considerable competition to sell

[23] *Wall Street Journal,* November 2, 1976.
[24] Ibid., November 22, 1976, p. 29.

enrichment services. In the decade of the 1990s, additional capacity will of course be needed, but there is no apparent reason to predict a shortage or lack of firms willing to enter the enrichment business.

Full "Privatization"

From an economic point of view, the sale of the existing government-owned plants makes great sense. If each of the three plants were sold to private groups and at the same time other firms were encouraged to build new facilities, a private market would be created. This market would be no more concentrated than that of aluminum or copper and could be expected to perform in a similar manner. That is, there would be market incentives for innovation and cost reduction for the existing plants as well as for the new facilities. In addition, the sale of the existing plants would assure the private sector that the government was serious about turning over enrichment to the market.

Selling the existing plants is made difficult because of the existing contracts that the government has with domestic utilities and foreign customers. About half are requirement contracts with prices based on "cost recovery" that probably cannot be increased, even under Carter's proposed commercial pricing policy. While ERDA believes that fixed commitment contracts (the other half) can be increased, they were obtained on the basis of discount pricing from requirement contracts.[25] Therefore, to what extent Carter or a private enricher can increase prices is problematical. As long as prices cover variable costs of the plants, however, these commitments would have the impact of lowering the amount that private investors would pay for the plants. Only if prices were below variable cost, which does not appear to be true, would the private sector be unwilling to pay anything for the plants. But these contracts may make the value of the plants so low that Congress would be unwilling to agree to a sale.

Since the relatively low transportation costs of enriched uranium make the market worldwide, the sale of the plants together with the construction of additional private and publicly owned facilities around the world should lead to a workably competitive industry by the late 1980s. If each of the four 2.2-million-SWU centrifuge modules that the government is building at Portsmouth and the three gaseous diffusion plants were sold to different enterprises, there would be seven domestic producers, plus several foreign producers. There should be no problem with a lack of competition.

At the time the government-owned plants were sold, it might be wise if the government stockpile of enriched uranium were also sold. The existence of a large privately owned and held stockpile of separative work could reduce any power that private enrichers might have to exploit any lack of competition in the initial period. Selling the stockpile, however, would reduce what the private

---

[25]*Nuclear Industry,* August 1977, p. 15.

sector would be willing to pay for the existing plants. It should be recognized that, whether the government sells the stockpile directly to utilities or to speculators, the impact will be the same on the sale price of the enrichment plants.

The sale of the government-owned facilities could be handled in a variety of ways. An unrestricted sale to the highest bidder would produce the greatest return for the government and the taxpayer. Any potential monopoly profits that were foreseen by the private sector would be reflected in the bid prices. Alternatively, limits could be put on the purchasers' rights to raise prices, build add-on facilities, build other new facilities, or make significant changes in toll contracts. Any such restrictions would limit not only the power of the purchaser but would also limit the amount the private sector would pay for the plants.

Because there are economies of scale up to very large sizes for the gaseous diffusion process, there would be incentives for a single private purchaser to acquire all the existing plants as a unit and to expand them rather than building new facilities. As a minimum, to promote the maximum number of competitors the existing plants should be sold to three separative enterprises. The Portsmouth centrifuge modules should be sold to four other firms.

Price constraints do not seem desirable. Regulation of prices in other industries has been singularly unsuccessful and has often resulted in considerable waste. If the purchasers of the government plants raise their charges for enrichment to reflect economic realities, that will lead to more conservation and a more appropriate balance of energy resources. Provided that competition exists—and seven enterprises is enough to ensure that—there need be no controls. Even if competition were temporarily weak, the absence of controls together with a policy of free entry would facilitate new construction and additional competition.

There are several precedents for the sale of major government enterprises to the private sector. The two biggest and most recent are the sales after World War II of government-constructed aluminum plants and synthetic rubber plants. In both cases the results have been successful, and the industries have become workably competitive. Industry performance in terms of innovation, relative prices, and profits has been good—in fact, generally better than before the sale.[26]

The arguments against the sale of the government plants are concerned with national security and the financial return to the government. The national security issue revolves around the protection of highly secret technology, the availability of capacity to enrich uranium highly for military purposes, and the proliferation of enrichment facilities abroad. In general, government officials who have examined the national security issue have concluded that the sale of the plants would have little effect on security.

---

[26]Robert Solo, ''Research and Development in the Synthetic Rubber Industry,'' *Quarterly Journal of Economics,* vol. 68 (February 1954).

The 1969 White House task force specifically considered the national security implications of the sale of the existing plants to the private sector. With the partial dissent of the Atomic Energy Commission, it concluded that adequate provisions could be made to ensure that there would be prompt availability of highly enriched uranium for national defense purposes and that the U.S. government could control the dissemination of enrichment technology and prevent diversion of enriched uranium to unauthorized uses, although some additional risk would exist. The task force also stated that, "while there might be some additional incentive to other countries to develop their own enrichment facilities, the risk does not appear sufficient to be a compelling reason against transfer of enrichment plants to private ownership."[27]

In the past the discussion of the sale of the enrichment plants has involved considerable controversy over a second issue: the financial desirability of the sale. Fears have been expressed that the plants would be sold at much less than they were worth, with their "value" based on government investment costs minus depreciation. But of course the value of an asset depends on its future earning power and not on its past construction costs. Since private owners will be interested in maximizing future earnings from the plants and since public ownership often results in practices which do not maximize earnings, these plants are worth more sold to the private sector than held in government hands. In other words, the taxpayer will be better off if the plants are sold than if they are kept under government operation. As for the loss of slight savings due to operating the plants as a unit, the Atomic Energy Commission has estimated these savings to be in the order of only 1 percent.[28]

As mentioned above, the existing enrichment contracts may reduce the economic viability of the gaseous diffusion plants to a low level. Private investors may be unwilling to bid much for the plants when enrichment services will have to be sold to existing customers until their contracts expire at prices below long-run costs. In the past Congress has been very sensitive to the issue of the value of the plants. Unless sentiment has changed significantly since the late 1960s, Congress will expect bids to equal or exceed the depreciated book value of the gaseous diffusion plants, plus an allowance for the amount invested in the capacity expansion program. Depending on the timing of the sale, Congress could expect over $2 billion for the three plants. It is unlikely that investors would be willing to put that much into what may be an obsolete technology which is tied into long-term contracts with utilities at low prices. Unless the government and the public recognize that the value of the plants is independent of historical costs and is simply a function of expected future earnings, there is no possibility of selling the gaseous diffusion plants.

Nevertheless, there is much to be gained not only from permitting the

[27]*Report of Task Force on Uranium Enrichment Facilities* (unclassified version) made available to the Joint Committee on Atomic Energy, June 29, 1960, p. 22.

[28]*AEC Gaseous Diffusion Plant Operations,* ORO-658, February 1968, p. 23.

private sector to build new enrichment facilities but also from selling the existing plants. Sale of the gaseous diffusion plants would assure the private sector that the government is sincere about turning over the enrichment industry to the private industry. It would therefore reduce the need for government guarantees of the type that UEA wanted. With the government out of the picture, there would be an assured market and a working technology to supply that market. The questions of when to build additional capacity, and how much to build, would become normal business decisions subject to the same risk such decisions face in other industries. Probably the only guarantee that the government need offer would be insurance against the government itself preventing the expansion of nuclear power.

Finally, turning the industry over to the private sector would ensure that the business would be operated efficiently without subsidy and on a strictly commercial basis. For the first time fossil fuels would compete with atomic energy on an equal footing.

6

Policy Prescriptions

Notwithstanding the efforts of environmentalists, the demand for energy can be expected to grow steadily throughout the rest of this century and into the next. While economies are possible, the need for additional electricity to light and heat new homes, offices, and factories cannot seriously be disputed. Fossil fuels, especially coal and oil, will be plentiful for the rest of this century and well into the next, but the cost of extracting them and converting them into clean, nonpolluting electricity will probably continue to rise. Therefore, even though President Carter has asserted that atomic energy is to be played down and depended on only as a last resort, there seems little doubt that its importance will grow.[1] In June 1977, James Schlesinger, speaking for the administration, explained that "the last resort" meant that by the year 2000 there would be 380 nuclear power plants in the United States, compared with the 66 then operating.[2]

There may be a breeder reactor in our future, but it seems far enough away that there will have to be a considerable expansion in uranium-enrichment capacity to supply the many light water reactors that will be constructed in the next couple of decades. As explained earlier, by 1985 or shortly thereafter, the demand for separative work will exceed the supply now planned for that period. Thus, there is considerable room for new enterprises to build and commercially operate enrichment facilities. Abroad, several enrichment projects are planned. In this country existing enrichment capacity is already committed and new facilities will be necessary by the second half of the 1980s.

The government therefore is faced with four alternatives: it may keep the ownership of enrichment facilities in its own hands, expanding capacity as necessary; it may establish a government-owned corporation to operate the plants in a manner similar to a private corporation and to construct any new

---

[1] *Nuclear Industry,* September 1976, p. 27.
[2] *Wall Street Journal,* October 19, 1977.

facilities; it may keep the existing plants and build the Portsmouth add-on facilities but permit and encourage the private sector to construct new plants; or it may decide that the whole industry, including the existing plants, should be turned over to the private sector.

From an economic point of view, the desirability of the various options appears to be in the order listed above, with the most desirable being the last. Maintaining the status quo implies continuing to underprice resources devoted to nuclear power and thus encouraging excessive construction of atomic power plants and excessive use of electricity. It also means that research and development will not be conducted on a strictly commercial basis, with the result that scarce resources will be used less effectively than they could be.

Although establishing a government corporation would improve incentives to economize, it would inevitably mean that scarce capital would be underpriced because of the government's advantageous position in the capital market. It might easily lead to congressional and executive pressure to favor certain groups.

The third alternative, which was reflected in the Nuclear Fuel Assurance Act proposed by the Ford administration, would have the benefit of inducing the private sector to enter the market and construct new enrichment facilities. While this would provide incentives for innovation and cost minimization, the existence of three plants in government hands would imply less than a full commitment to "privatization." These government plants could still be operated in a noncommercial manner, and some separative work could continue to be underpriced. So long as the government does not appear to be fully committed to turning over the enrichment business to the private sector, commercial firms are going to demand and need some government guarantees to invest billions in a new industry. With complete privatization the risks involved in investing in the enrichment business would be similar to those in other industries with large capital investments such as steel, aluminum, and petroleum refining.

The recent decision by the Carter administration to expand the Portsmouth plant will slow down any transfer of the industry to the private sector. This is because even if all other increments of capacity would be built by private enterprise, the addition of the 8.8-million-SWU capacity to the existing facilities postpones for a few years the need for more capacity.

Selling the government plants would eliminate the need to provide most guarantees, especially if the government stockpile of enriched uranium were sold at the same time. (It might be wise for the government to offer insurance against regulations foreclosing the nuclear power industry. That is, the government could agree to indemnify private investors if due to government controls the nuclear power industry were kept from growing.) If the private stockpile were held by investors, private firms could be sure that even if the new enrichment facilities being constructed were delayed, they could purchase enough enriched uranium on the open market to meet any contractual obligations.

The advantages of "privatization" are that it would eliminate the bias in favor of nuclear power stemming from the underpricing of separative work, provide strong market incentives to innovate and to reduce costs, and reduce if not eliminate political considerations in nuclear power. Turning the enrichment industry over to the private sector would no doubt lead to a considerable increase in the price for separative work in the short run, but with market incentives and competition it is quite possible that privatization would mean lower prices, as well as lower costs, in the long run, than would continued government ownership.

Appendix

The Plutonium Recycle Question

As the text pointed out, plutonium recyling would reduce the need for separative work, but there are two major considerations that must be taken into account in evaluating the spent fuel recycle question. First, is it economically worthwhile to recycle the spent fuel? Second, are there security or environmental risks that would indicate that it should not be done, even if it is economically worthwhile?

According to ERDA figures, recycling uranium and plutonium would save about 7 percent of the total fuel cycle costs, excluding reactors.[1] On the other hand according to engineers from the Electric Power Research Institute, "Eventually, with the reprocessing of spent fuel, a considerable credit for uranium and plutonium content may be realized. However, it is unlikely that this credit will be significantly larger than the costs of interim storage and eventually reprocessing, and in any event the present worth of these credits is small and may be safely neglected."[2] The rather negative conclusions of these engineers are based on the existing moratorium on reprocessing, which means that if approval were given promptly to go ahead, large-scale reprocessing would not be viable until the mid-1980s at the earliest.

President Carter announced on April 7, 1977, that "we will defer indefinitely the commercial reprocessing and recycling of the plutonium produced in U.S. nuclear power programs." The plant at Barnwell, South Carolina, for instance (see below), will receive neither federal encouragement nor funding for its completion as a reprocessing facility.[3] In October the President proposed that the federal government buy and store the used radioactive fuel from

---

[1] ERDA, *Report: Light Water Reactor Fuel Reprocessing and Recycling*, July 1977, p. 2.3.

[2] M. Levenson and E. Zebroski, "The Nuclear Fuel Cycle," *Annual Review of Energy*, vol. 1, edited by Jack M. Hollander (Palo Alto, Calif.: Annual Reviews, Inc., 1976), pp. 101-30.

[3] Presidential Statement, April 7, 1977, published in *Weekly Compilation of Presidential Documents*, April 11, 1977, vol. 13–Number 15, p. 503.

commercial nuclear power plants, both foreign and domestic.[4] One objective of this program is to reduce incentives for foreign nations to develop reprocessing facilities.

The Carter administration, however, has not banned reprocessing, only deferred it. In fact, James Schlesinger, secretary of the Department of Energy, said in October 1977, that "reprocessing is probably an indispensable step towards the ultimate solution of the waste problem."[5]

The technology of spent fuel recycling is well known and at least ten very large-scale reprocessing plants have been built, four of them in the United States.[6] The recent history of reprocessing within the United States, however, has not been good. Nuclear Fuel Service, Inc., the only company ever to operate a commercial nuclear fuel reprocessing plant, is trying to abandon its efforts, as a consequence of changes in regulatory requirements and other factors.[7] Under pressure from environmentalists, the Nuclear Regulatory Commission has been moving to impose more stringent requirements on reprocessing plants, and it has yet to license one. Besides the Nuclear Fuel Service plant, which has not operated since 1972 when the Atomic Energy Commission shut it down, Allied Chemical Corporation has poured more than $250 million into the plant in Barnwell, South Carolina, which is still two years from production. After it is finished, if it ever is, nearly half a billion dollars will have to be invested in auxiliary facilities to convert the plutonium-bearing liquid to a solid and to convert the high-level radioactive wastes into solids that can be stored. Finally, General Electric has put $64 million into a reprocessing plant using a different technique, which proved unsuccessful.

But the problems are more than just financial and technological; they involve environmental and safety considerations. The waste material from recycling is highly radioactive, and plutonium is extremely toxic. Leaks, spills, and other accidents in moving the spent fuel from the reactor to the reprocessing plant, in moving it at the reprocessing plant, or even in shipping the plutonium to a fuel fabrication plant could produce deaths, injuries, and disease in populations exposed to the wastes. Environmental groups are opposing the licensing of any such reprocessing plant on the grounds that such plants generate great risks for the population and the environment.

In addition, while natural uranium and even uranium enriched to 4 percent U_{235} cannot be used to make bombs, plutonium can. Thus, although a normal reactor site or an enriching plant is not a useful target for a terrorist group, the theft of plutonium makes a credible threat against society.

These dangers can be overstated, however. Given its highly radioactive

---

[4]*Wall Street Journal,* October 18, 1977, p. 18.

[5]Ibid., p. 18.

[6]Levenson and Zebroski, "The Nuclear Fuel Cycle," *Annual Review,* p. 663.

[7]Tim Metz, "Getty Oil's NFS Unit Weighing Pullout from Nuclear Fuel Reprocessing Project," *Wall Street Journal,* July 15, 1976, p. 9.

nature and its extreme toxicity, plutonium is extremely dangerous to handle; the inhalation of one ten-thousandth of a gram apparently is sufficient to cause lung cancer.[8] To build any type of weapon with such material requires expensive and sophisticated equipment and highly trained personnel. This would be beyond the capabilities of almost any terrorist organization and would almost surely have to be the work of a government. Greater danger would result from a terrorist organization capturing or stealing one of the more than 80,000 nuclear weapons deployed around the world, a possibility that is not affected by reprocessing of fuel.

A more plausible problem would be a terrorist's threat of exploding or scattering plutonium over an inhabited area. In such a situation, thousands could be affected, but it would take the scattering of considerable quantities of plutonium to result in quick deaths, although small quantities could result in the development of lung cancer and eventual death in the exposed population.[9] It would be easier and cheaper for terrorists to use chemical or biological agents to threaten large population centers than to attempt to scatter nuclear wastes.

Nevertheless, there is little danger from spent fuel unless it is reprocessed. No terrorist group would find the waste fuel now produced by enrichment plants a particularly effective threat against the public. Not reprocessing might raise power costs between now and the year 2000 by $12 billion, or about 2.5 percent,[10] but this additional outlay could be considered the premium necessary to reduce the risks from plutonium to negligible levels. Moreover, it should be remembered that stored waste fuel will remain as a potential source of energy if at any time in the future reprocessing is deemed advisable.

Even though reprocessing has been deferred, other countries will undoubtedly go ahead with it. The Germans have already agreed to build a complete nuclear fuel cycle for Brazil, including a reprocessing plant. The French have such a plant and have agreed to build one in Pakistan. No doubt the Indians also have one. Further proliferation of reprocessing plants seems inevitable whether the United States permits reprocessing or not. The worldwide nuclear weapon proliferation problem will be little affected by the government decision on American reprocessing; any objective analysis of the risk of terrorist activities suggests that those problems will be significantly greater abroad than here and that American reprocessing will make little difference.

There are a number of real advantages to reprocessing the wastes from nuclear power plants. It would reduce the physical bulk of waste materials that must be stored indefinitely and thus simplify the waste management problem. It would also reduce the need for yellowcake and preserve existing supplies of uranium by recycling uranium and by substituting plutonium for enriched

---

[8] ERDA, *Report: Light Water Reactor Fuel Reprocessing and Recycling,* p. 10.15.
[9] Ibid., pp. 10.15-10.16.
[10] Ibid., pp. 5.19 and 5.48.

uranium. This would also reduce the need for enrichment services. The Edison Electric Institute has estimated that reprocessing could reduce the need for virgin uranium by 33 percent, but one major utility has indicated that the savings would be only 16 percent if plutonium could not be used.[11] ERDA estimated that prompt introduction of recycling could reduce both uranium ore and enrichment requirements by 25 percent by the year 2000.[12] Edison Electric Institute has also estimated that utilities could save from 5 to 22 percent on fuel cycle costs by reprocessing, though this does not include any safeguard costs.[13]

It is unlikely that the private sector will attempt to develop any further recycling. Given President Carter's decision to defer recycling, the probability of being licensed by the Nuclear Regulatory Commission is so low that no company would go forward on its own. Moreover, with the administration prepared to take nuclear wastes off the hands of utilities, the demand by power companies is considerably allayed. Eventually, as Secretary Schlesinger has indicated, the Carter administration may support reprocessing. While the benefits from reprocessing are not great, they appear to be greater than the costs, including safeguard costs.

---

[11]*Wall Street Journal,* February 17, 1976, pp. 1 and 18.

[12]ERDA, *Report: Light Water Reactor Fuel Reprocessing and Recycling,* p. 2.2.

[13]*Wall Street Journal,* February 17, 1976, pp. 1 and 18.